D1266961

The Proof is in the Pudding

Steven G. Krantz

The Proof is in the Pudding

The Changing Nature of Mathematical Proof

 Springer

Steven G. Krantz
Department of Mathematics
Washington University in St. Louis
St. Louis, MO 63130
USA
sk@math.wustl.edu

ISBN 978-0-387-48908-7   e-ISBN 978-0-387-48744-1
DOI 10.1007/978-0-387-48744-1
Springer New York Dordrecht Heidelberg London

Library of Congress Control Number: 2011928557

Mathematics Subject Classification (2010): 01-XX, 03-XX, 65-XX, 97A30

© Springer Science+Business Media, LLC 2011
All rights reserved. This work may not be translated or copied in whole or in part without the written permission of
the publisher (Springer Science+Business Media, LLC, 233 Spring Street, New York, NY 10013, USA), except for
brief excerpts in connection with reviews or scholarly analysis. Use in connection with any form of information storage
and retrieval, electronic adaptation, computer software, or by similar or dissimilar methodology now known or hereafter
developed is forbidden.
The use in this publication of trade names, trademarks, service marks, and similar terms, even if they are not identified
as such, is not to be taken as an expression of opinion as to whether or not they are subject to proprietary rights.

Printed on acid-free paper

Springer is part of Springer Science+Business Media (www.springer.com)

ACC LIBRARY SERVICES AUSTIN, TX

To Jerry Lyons, mentor and friend

Preface

The title of this book is not entirely frivolous. There are many who will claim that the correct aphorism is "The proof of the pudding is in the eating."—that it makes no sense to say, "The proof is in the pudding." Yet people say it all the time, and the intended meaning is usually clear. So it is with mathematical proof. A *proof* in mathematics is a psychological device for convincing some person, or some audience, that a certain mathematical assertion is true. The structure, and the language used, in formulating such a proof will be a product of the person creating it; but it also must be tailored to the audience that will be receiving it and evaluating it. Thus there is no "unique" or "right" or "best" proof of any given result. A proof is part of a situational ethic: situations change, mathematical values and standards develop and evolve, and thus the very *way* that mathematics is done will alter and grow.

This is a book about the changing and growing nature of mathematical proof. In the earliest days of mathematics, "truths" were established heuristically and/or empirically. There was a heavy emphasis on calculation. There was almost no theory, no formalism, and there was little in the way of mathematical notation as we know it today. Those who wanted to consider mathematical questions were thereby hindered: they had difficulty expressing their thoughts. They had particular trouble formulating general statements about mathematical ideas. Thus it was virtually impossible for them to state theorems and prove them.

Although there are some indications of proofs even on ancient Babylonian tablets (such as Plimpton 322) from 1800 BCE, it seems that it is in ancient Greece that we find the identifiable provenance of the concept of proof. The earliest mathematical tablets contained numbers and elementary calculations. Because of the paucity of texts that have survived, we do not know how it came about that someone decided that some of these mathematical procedures required *logical justification*. And we really do not know how the formal concept of proof evolved. The *Republic* of Plato contains a clear articulation of the proof concept. The *Physics* of Aristotle not only discusses proofs, but treats minute distinctions of proof methodology. Many of the ancient Greeks, including Eudoxus, Theaetetus, Thales, Euclid, and Pythagoras, either used proofs or referred to proofs. Protagoras was a sophist, whose work was recognized by Plato. His *Antilogies* were tightly knit, rigorous arguments that could be thought of as the germs of proofs.

It is acknowledged that Euclid was the first to systematically use precise definitions, axioms, and strict rules of logic, and to carefully enunciate and *prove* every statement (i.e., every theorem). Euclid's formalism, as well as his methodology, has become the model—even to the present day—for establishing mathematical facts.

What is interesting is that a mathematical statement of fact is a freestanding entity with intrinsic merit and value. But a proof is a device of communication. The creator or discoverer of this new mathematical result wants others to believe it and accept it. In the physical sciences—chemistry, biology, or physics for example—the method for achieving this end is the *reproducible experiment*.[1] For the mathematician, the reproducible experiment is a proof that others can read and understand and validate.

The idea of "proof" appears in many aspects of life other than mathematics. In the courtroom, a lawyer (either for the prosecution or the defense) must establish a case by means of an accepted version of proof. For a criminal case this is "beyond a reasonable doubt," while for a civil case it is "the preponderance of evidence shows." Neither of these is mathematical proof, or anything like it. The real world has no formal definitions and no axioms; there is no sense of establishing facts by strict exegesis. The lawyer uses logic—such as "the defendant is blind so he could not have driven to Topanga Canyon on the night of March 23," or "the defendant has no education and therefore could not have built the atomic bomb that was used to . . . "—but the principal tools are *facts*. The lawyer proves the case beyond a reasonable doubt by amassing an overwhelming preponderance of evidence in favor of that case.

At the same time, in ordinary, family-style parlance there is a notion of proof that is different from mathematical proof. A husband might say, "I believe that my wife is pregnant," while the wife may *know* that she is pregnant. Her pregnancy is not a permanent and immutable fact (like the Pythagorean theorem), but instead is a "temporary fact" that will be false after several months. So, in this context, the concept of truth has a different meaning from the one used in mathematics, and the means of verification of a truth are also rather different. What we are really seeing here is the difference between knowledge and belief—something that never plays a formal role in mathematics.

In modern society it takes far less "proof" to convict someone of speeding than to get a murder conviction. But, ironically, it seems to take even less evidence to justify waging war.[2] A great panorama of opinions about the modern concept of proof—in many different contexts—may be found in [NCBI]. It has been said—see [MCI]—that in some areas of mathematics (such as low-dimensional topology) a proof can be conveyed by a combination of gestures.

The French mathematician Jean Leray (1906–1998) perhaps sums up the value system of the modern mathematician:

> . . . all the different fields of mathematics are as inseparable as the different parts of a living organism; as a living organism mathematics has to be permanently recreated; each generation must reconstruct it wider, larger and more beautiful. The death of mathematical research would be the death of mathematical thinking which constitutes the structure of scientific language itself and by consequence the death of our scientific civilization. Therefore we must transmit to our children strength of character, moral values and drive towards an endeavouring life.

---

[1] More precisely, it is the reproducible experiment *with control*. For the careful scientist compares the results of his/her experiment with some standard or norm. That is the means of evaluating the result.

[2] See [SWD] for the provenance of these ideas. Fascinating related articles are [ASC], [MCI]. An entire issue of the *Philosophical Transactions of the Royal Society* in the fall of 2005 is devoted to articles of this type, all deriving from a meeting in Britain to discuss issues such as those treated in this book. See [PTRS].

What Leray is telling us is that mathematical ideas travel well and stand up under the test of time, because we have such a rigorous and well-tested standard for formulating and recording the ideas. It is a grand tradition, and one well worth preserving.

There is a human aspect to proof that cannot be ignored. The acceptance of a new mathematical truth is a sociological process. It is something that takes place in the mathematical community. It involves understanding, internalization, rethinking, and discussion. And even the most eminent mathematicians make mistakes, and announce new results that in fact they *do not* know how to prove. In 1879, A. Kempe published a proof of the four-color theorem that stood for eleven years before P. Heawood found a fatal error in the work. The first joint work of G. H. Hardy and J. E. Littlewood was announced at the June, 1911 meeting of the London Mathematical Society. The result was never published because they later discovered that their proof was incorrect. A. L. Cauchy, G. Lamé, and E. E. Kummer all thought at one point or another in their careers that they had proved Fermat's last theorem; they were all mistaken. H. Rademacher thought in 1945 that he had disproved the Riemann hypothesis; his achievement was even written up in *Time Magazine*. He later had to withdraw the claim because C. L. Siegel found an error. Considerable time is spent here in this book exploring the social workings of the mathematical discipline, and how the interactions of different mathematicians and different mathematical cultures serve to shape the subject. Mathematicians' errors are corrected, not by formal mathematical logic, but by other mathematicians. This is a seminal point about this discipline.[3]

The early twentieth century saw L. E. J. Brouwer's dramatic proof of his fixed-point theorem followed by his wholesale rejection of proof by contradiction (at least in the context of existence proofs—which is precisely what the proof of his fixed-point theorem is an instance of) and his creation of the intuitionist movement. This program was later taken up by Errett Bishop, and his *Foundations of Constructive Analysis* (written in 1967) has made quite a mark (see also the revised version, written jointly with Douglas Bridges, published in 1985). These ideas are of particular interest to the theoretical computer scientist, for proof by contradiction has questionable meaning in computer science (this despite the fact that Alan Turing cracked the Enigma code by applying ideas of proof by contradiction in the context of computing machines).

In the past thirty years or so it has come about that we have rethought, and reinvented, and decisively amplified our concept of proof. Computers have played a strong and dynamic role in this reorientation of the discipline. A computer can make hundreds of millions of calculations in a second. This opens up possibilities for trying things, calculating, and visualizing things that were unthinkable fifty years ago. It should be borne in mind that mathematical thinking involves concepts and reasoning, while a computer is a device for manipulating data, two quite different activities. It appears unlikely (see Roger Penrose's remarkable book *The Emperor's New Mind*) that a computer will ever be able to think and prove mathematical theorems in the way a human being performs these activities. Nonetheless, the computer can provide valuable information and insights. It can enable the user to see things that would not be envisioned otherwise. It is a valuable tool. We shall

---

[3]It is common for mathematicians to make errors. Probably every published mathematical paper has errors in it. The book [LEC] documents many important errors in the literature up to the year 1935.

definitely spend a good deal of time in this book pondering the role of the computer in modern human thought.

In trying to understand the role of the computer in mathematical life, it is perhaps worth drawing an analogy with history. Tycho Brahe (1546–1601) was one of the great astronomers of the Renaissance. Through painstaking scientific procedure, he recorded reams and reams of data about the motions of the planets. His gifted student Johannes Kepler (1571–1630) was anxious to get his hands on Brahe's data because he had ideas about formulating mathematical laws about the motions of the planets. But Brahe and Kepler were both strong-willed men. They did not see eye-to-eye on many things. And Brahe feared that Kepler would use his data to confirm the Copernican theory about the solar system (namely that the *sun*, not the earth, was the center of the system—a notion that ran counter to Christian religious dogma). As a result, during Brahe's lifetime Kepler did not have access to Brahe's numbers.

But providence intervened in a strange way. Tycho Brahe had been given an island by his sponsor on which to build and run his observatory. As a result, he was obliged to attend certain social functions—just to show his appreciation and to report on his progress. At one such function, Brahe drank an excessive amount of beer, his bladder burst, and he died. Kepler was then able to negotiate with Brahe's family to get the data that he so desperately needed. And thus the course of scientific history was forever altered.

Kepler did *not* use deductive thinking or reasoning, or the axiomatic method, or the strategy of mathematical proof to derive his three laws of planetary motion. Instead he simply stared at the hundreds of pages of planetary data that Brahe had provided, and he performed numerous calculations.

At around this same time John Napier (1550–1617) was developing his theory of logarithms. These are terrific calculational tools, which would have simplified Kepler's task immensely. But Kepler could not understand the derivation of logarithms and refused to use them. He did everything the hard way. Imagine what Kepler could have done with a computer!—but he probably would have refused to use one just because he would not have understood how the central processing unit (CPU) worked.

In any event, we tell here of Kepler and Napier because the situation is perhaps a harbinger of modern agonizing over the use of computers in mathematics. There are those who argue that the computer can enable us to see things—both calculationally and visually—that we could not see before. And there are those who say that all those calculations are well and good, but they do not constitute a mathematical proof. Nonetheless it seems that the first can inform the second, and a productive symbiosis can be created. We shall discuss these matters in detail as the book develops.

Now let us return to our consideration of changes that have come about in mathematics in the past thirty years, in part because of the advent of high-speed digital computers. Here is a litany of some of the components of this process:

(a) In 1974, Appel and Haken [APH1] announced a proof of the four-color conjecture. This is the question of how many colors are needed to color any map, so that adjacent countries are colored differently. Their proof used 1200 hours of computer time on a supercomputer at the University of Illinois. Mathematicians found this event puzzling

because this "proof" was not something that anyone could study or check. Or understand. To this day there does not exist a proof of the four-color theorem that can be
read and checked by humans.

(b)   Over time, people have become more and more comfortable with the use of computers
in proofs. In its early days, the theory of wavelets (for example) depended on the
estimation of a certain constant—something that could be done only with a computer.
De Branges's original proof of the Bieberbach conjecture [DEB2] seemed to depend
on a result from special function theory that could be verified only with the aid of a
computer (it was later discovered to be a result of Askey and Gasper that was proved
in the traditional manner).

(c)   The evolution of new teaching tools such as the software The Geometer's
Sketchpad has suggested to many—including Fields Medalist William Thurston—
that traditional proofs may be set aside in favor of experimentation, that is, testing of
thousands or millions of examples, on the computer.

Thus the use of the computer has truly reoriented our view of what a proof might
comprise. Again, the point is to convince someone else that something is true. There are
evidently many different means of doing this.

Perhaps more interesting are some of the new social trends in mathematics and the
resulting construction of nonstandard proofs (we shall discuss these in detail in the text that
follows):

(a)   One of the great efforts of twentieth-century mathematics has been the classification of
the finite simple groups. Daniel Gorenstein, of Rutgers University, was in some sense
the lightning rod who orchestrated the effort. It is now considered to be complete.
What is remarkable is that this is a single theorem that is the aggregate effort of many
hundreds of mathematicians. The "proof" is in fact the union of hundreds of papers
and tracts spanning more than 150 years. At the moment this proof comprises over
10,000 pages. It is still being organized and distilled today. The final "proof for the
record" will consist of several volumes, and it is not clear that the living experts will
survive long enough to see the fruition of this work.

(b)   Thomas Hales's resolution of the Kepler sphere-packing problem uses a great deal of
computer calculation, much as with the four-color theorem. It is particularly interesting
that his proof supplants the earlier proof of Wu-Yi Hsiang that relied on spherical
trigonometry and *no computer calculation*. Hales allows that his "proof" cannot be
checked in the usual fashion. He has organized a worldwide group of volunteers called
FlySpeck to engage in a checking procedure for his computer-based arguments.

(c)   Grisha Perelman's "proof" of the Poincaré conjecture and the geometrization program
of Thurston are currently in everyone's focus. In 2003, Perelman wrote three papers that
describe how to use Richard Hamilton's theory of Ricci flows to carry out Thurston's
idea (called the "geometrization program") of breaking up a 3-manifold into fundamental geometric pieces. One very important consequence of this result would be a
proof of the important Poincaré conjecture. Although Perelman's papers are vague
and incomplete, they are full of imaginative and deep geometric ideas. This work
set off a storm of activity and speculation about how the program might be assessed

and validated. There have been huge efforts by John Lott and Bruce Kleiner (at the University of Michigan) and Gang Tian (Princeton) and John Morgan (Columbia) to complete the Hamilton/Perelman program and produce a bona fide, recorded proof that others can study and verify.

(d)   In fact, Thurston's geometrization program is a tale in itself. He announced in the early 1980s that he had a result on the structure of 3-manifolds, at least for certain important subclasses of the manifolds, and he knew how to prove it. The classical Poincaré conjecture would be an easy corollary of Thurston's geometrization program. He wrote an extensive set of notes [THU3]—of book length—and these were made available to the world by the Princeton mathematics department. For a nominal fee, the department would send a copy to anyone who requested it. These notes, entitled *The Geometry and Topology of Three-Manifolds* [THU3], were extremely exciting and enticing. But the notes, for all the wealth of good mathematics that they contained, were written in a rather informal style. They were difficult to assess and evaluate.

The purpose of this book is to explore all the ideas and developments outlined above. Along the way, we are able to acquaint the reader with the culture of mathematics: who mathematicians are, what they care about, and what they do. We also give indications of why mathematics is important, and why it is having such a powerful influence in the world today. We hope that by reading this book the reader will become acquainted with, and perhaps charmed by, the glory of this ancient subject, and will realize that there is so much more to learn.

December, 2010   Steven G. Krantz
                                                                      St. Louis, Missouri

Contents

Acknowledgments

One of the pleasures of the writing life is getting criticism and input from other scholars. I thank Jesse Alama, David H. Bailey, John Bland, Jonathan Borwein, Robert Burckel, David Collins, E. Brian Davies, Keith Devlin, Ed Dunne, Michael Eastwood, Jerry Folland, H. Gopalkrishna Gadiyar, Jeremy Gray, Jeff C. Lagarias, Barry Mazur, Robert Strichartz, Eric Tressler, James Walker, Russ Woodroofe, and Doron Zeilberger for careful readings of my drafts and for contributing much wisdom and useful information. Robert Burckel and David Collins each gave my draft an extraordinarily incisive and detailed reading and contributed many useful ideas and corrections. I thank Ed Dunne of the American Mathematical Society for many insights and for suggesting the topic of this book. Finally, I thank Sidney Harris for permission to use six of his cartoons.

Ann Kostant of Birkhäuser/Springer was, as always, a proactive and supportive editor. She originally invited me to write a book for the Copernicus series, and gave terrific advice and encouragement during its development. Edwin Beschler served as editor extraordinaire, and helped me sharpen the prose and enliven the exposition. David Kramer was, as always, a splendid copyeditor. I am proud of the result. Finally, I thank Randi D. Ruden for all her help and support during the writing of this book.

1

What Is a Proof and Why?

The proof of the pudding is in the eating.

—Miguel Cervantes

In mathematics there are no true controversies.

—Carl Friedrich Gauss

Logic is the art of going wrong with confidence.

—Anonymous

To test man, the proofs shift.

—Robert Browning

Newton was a most fortunate man because there is just one universe and Newton had discovered its laws.

—Pierre-Simon Laplace

The chief aim of all investigations of the external world should be to discover the rational order and harmony which has been imposed on it by God and which He revealed to us in the language of mathematics.

—Johannes Kepler

A terrier may not be able to define a rat, but a terrier knows a rat when he sees it.

—A. E. Housman

1.1 What Is a Mathematician?

A well-meaning mother was once heard telling her child that a mathematician is someone who does "scientific arithmetic." Others think that a mathematician is someone who spends all day hacking away at a computer.

Neither of these contentions is incorrect, but they do not begin to penetrate all that a mathematician really is. Paraphrasing mathematician/linguist Keith Devlin, we note that a mathematician is someone who:

Figure 1.1.

- observes and interprets phenomena
- analyzes scientific events and information
- formulates concepts
- generalizes concepts
- performs inductive reasoning
- performs analogical reasoning
- engages in trial and error (and evaluation)
- models ideas and phenomena
- formulates problems
- abstracts from problems
- solves problems
- uses computation to draw analytical conclusions
- makes deductions
- makes guesses
- proves theorems

And even this list is incomplete. A mathematician has to be a master of critical thinking, analysis, and deductive logic. These skills are robust, and can be applied in a large variety of situations—and in many different disciplines. Today, mathematical skills are being put to good use in medicine, physics, law, commerce, Internet design, engineering, chemistry, biological science, social science, anthropology, genetics, warfare, cryptography, plastic surgery, security analysis, data manipulation, computer science, and in many other disciplines and endeavors as well.

One of the astonishing and dramatic new uses of mathematics that has come about in the past twenty years is in finance. The work of Fischer Black of Harvard and Myron

Scholes of Stanford gave rise to the first-ever method for option pricing. This methodology is based on the theory of stochastic integrals—a part of abstract probability theory. As a result, investment firms all over the world now routinely employ Ph.D. mathematicians. When a measure theory course is taught in the math department—something that was formerly the exclusive province of graduate students in mathematics studying for the qualifying exams— we find that the class is unusually large, and most of the students are from economics and finance.

Another part of the modern world that has been strongly influenced by mathematics, and which employs a goodly number of mathematicians with advanced training, is genetics and the genome project. Most people do not realize that a strand of DNA can have billions of gene sites on it. Matching up genetic markers is *not* like matching up your socks; in fact things must be done probabilistically. A good deal of statistical theory is used. So these days many Ph.D. mathematicians work on the genome project.

The focus of this book is on the concept of *mathematical proof*. Although it is safe to say that most mathematical scientists do not spend the bulk of their time proving theorems,[1] it is nevertheless the case that *proof* is the *lingua franca* of mathematics. It is the web that holds the enterprise together. It is what makes the subject live on, and guarantees that mathematical ideas will have some immortality (see [CEL] for a philosophical consideration of the proof concept).

There is no other scientific or analytical discipline that uses proof as readily and routinely as does mathematics. This is the device that makes theoretical mathematics special: the carefully crafted path of inference, following strict analytical rules that leads inexorably to a particular conclusion. *Proof* is our device for establishing the absolute and irrevocable truth of statements in mathematics. This is the reason we can depend on the mathematics of Euclid 2300 years ago as readily as we believe in the mathematics that is done today. No other discipline can make such an assertion (but see Section 1.10).

This book will acquaint the reader with some mathematicians and what they do, using the concept of "proof" as a touchstone. Along the way, we will become acquainted with foibles and traits of particular mathematicians, and of the profession as a whole. It is an exciting journey, full of rewards and surprises.

1.2 The Concept of Proof

We begin this discussion with an inspiring quotation from mathematician Michael Atiyah (1929–) [ATI2]:

> We all know what we like in music, painting or poetry, but it is much harder to explain why we like it. The same is true in mathematics, which is, in part, an art form. We can identify a long list of desirable qualities: beauty, elegance,

---

[1] This is because a great many mathematical scientists do not work at universities. They work instead (for instance) for the National Security Agency (NSA), or the National Aeronautics and Space Administration (NASA), or Hughes Aircraft, or Lawrence Berkeley Labs, or Microsoft. It is in fact arguable that most mathematical scientists are *not* pedigreed mathematicians.

"YOU WANT PROOF? I'LL GIVE YOU PROOF!"

Figure 1.2.

importance, originality, usefulness, depth, breadth, brevity, simplicity, clarity. However, a single work can hardly embody them all; in fact, some are mutually incompatible. Just as different qualities are appropriate in sonatas, quartets or symphonies, so mathematical compositions of varying types require different treatment. Architecture also provides a useful analogy. A cathedral, palace or castle calls for a very different treatment from an office block or private home. A building appeals to us because it has the right mix of attractive qualities for its purpose, but in the end, our aesthetic response is instinctive and subjective. The best critics frequently disagree.

The tradition of mathematics is a long and glorious one. Along with philosophy, it is the oldest venue of human intellectual inquiry. It is in the nature of the human condition to want to understand the world around us, and mathematics is a natural vehicle for doing so. But, for the ancients, mathematics was also a subject that was beautiful and worthwhile in its own right, a scholarly pursuit possessing intrinsic merit and aesthetic appeal. Mathematics was worth studying for its own sake.

In its earliest days, mathematics was often bound up with practical questions. The Egyptians, as well as the Greeks, were concerned with surveying land. Refer to Figure 1.3. Questions of geometry and trigonometry were natural considerations. Triangles and rectangles arose in a natural way in this context, so early geometry concentrated on these constructs. Circles too were natural to consider—for the design of arenas and water tanks and

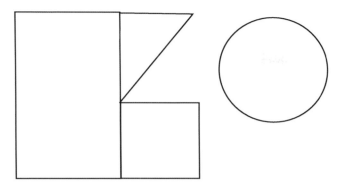

Figure 1.3. Surveying land.

other practical projects. So ancient geometry (and Euclid's axioms for geometry) discussed circles.

The earliest mathematics was phenomenological. If one could draw a plausible picture, then that was all the justification needed for a mathematical "fact." Sometimes one argued by analogy, or by invoking the gods. The notion that mathematical statements could be *proved* was not yet a developed idea. There was no standard for the concept of proof. The analytical structure, the "rules of the game," had not yet been created. If one ancient Egyptian were to say to another, "I don't understand why this mathematical statement is true. Please prove it," his request would have been incomprehensible. The concept of proof was not part of the working vocabulary of an ancient mathematician.

Well then, what is a proof? Heuristically, a proof is a rhetorical device for convincing someone else that a mathematical statement is true or valid. And how might one do this? A moment's thought suggests that a natural way to prove that something new (call it **B**) is true is to relate it to something old (call it **A**) that has already been accepted as true. In this way the concept arises of *deriving* a new result from an old result. See Figure 1.4. The next question is, "How was the old result verified?" Applying this regimen repeatedly, we find ourselves considering a sequence of logic as in Figure 1.5. But then one cannot help but ask, "Where does the chain begin?" And this is a fundamental issue.

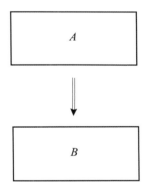

Figure 1.4. Logical derivation.

It will not do to say that the reasoning path has no beginning: that it extends infinitely far back into the mists of time because, if that were the case, it would undercut our thinking of what a proof should be. We are trying to justify new mathematical facts in terms of old mathematical facts. But if the inference regresses infinitely far back into the past, then we cannot actually ever grasp a basis or initial justification for our logic.

As a result of these questions, ancient mathematicians had to think hard about the nature of mathematical proof. Thales (640–546 BCE), Eudoxus (408–355 BCE), and Theaetetus of Athens (417–369 BCE) actually formulated theorems as formal enunciations of certain ideas that they wished to establish as facts or truths. Evidence exists that Thales proved some of these theorems in geometry (and these were later put into a broader context by Euclid). A theorem is the mathematician's formal enunciation of a fact or truth. But Eudoxus fell short in finding means to prove his theorems. His work had a distinctly practical bent, and he was particularly fond of calculations.

Euclid of Alexandria first formalized the way we now think about mathematics. Euclid had definitions and axioms and then theorems—in that order. There is no gainsaying the assertion that Euclid set the paradigm by which we have been practicing mathematics for 2300 years. This was mathematics done right. Now, following Euclid, in order to address the issue of the infinitely regressing chart of logic, we begin our studies by putting into place a set of *definitions* and a set of *axioms*.

What is a definition? A definition explains the meaning of a piece of terminology. There are analytical problems with even this simple idea; consider the first definition that we are going to formulate. Suppose that we wish to define a *rectangle*. This will be the first piece of terminology in our mathematical system. What words can we use to define it? Suppose that we define rectangle in terms of points and lines and planes. That begs the questions: What is a point? What is a line? What is a plane?

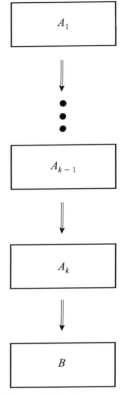

Figure 1.5. A sequence of logical steps.

Thus we see that our *first* definition(s) must be formulated in terms of commonly accepted words that require no further explanation. Aristotle (384–322 BCE) insisted that a definition must describe the concept being defined in terms of other concepts already known. This is often quite difficult. As an example, Euclid defined a *point* to be that which has no part. Thus he used words *outside of mathematics*, words that are a commonly accepted part of everyday jargon, to explain the precise mathematical notion of "point."[2] Once "point" is defined, then one can use that term in later definitions. And one can then use everyday language that does not require further explication. This is how we build up a system of definitions.

The definitions give us then a language for doing mathematics. We formulate our results, or *theorems*, using the words that have been established in the definitions. But wait, we are

---

[2]It is quite common among those who study the foundations of mathematics to refer to terms that are defined in nonmathematical language—that is, which cannot be defined in terms of other mathematical terms—as *undefined terms*. The concept of "set," which is discussed elsewhere in this book, is an undefined term. So is "point."

Figure 1.6.

not yet ready for theorems, because we have to lay cornerstones upon which our inference can develop. This is the purpose of axioms.

What is an axiom? An axiom[3] (or postulate[4]) is a mathematical statement of fact, formulated using terminology that has been defined in the definitions, that is taken to be self-evident. An axiom embodies a crisp, clean mathematical assertion. One does not *prove* an axiom. One takes an axiom to be given, and to be so obvious and plausible that no proof is required.

Axioms can also be used to explain primitives. These are ideas that are at the basis of the subject, and whose properties are considered to be self-explanatory or self-evident. Again, one cannot verify the assertions in an axiom. They are presented for the reader's enjoyment, with the understanding that they will be used in what follows to prove mathematical results.

One of the most famous axioms in all of mathematics is the *parallel postulate* of Euclid. The parallel postulate (in Playfair's formulation)[5] asserts that if P is a point, and if ℓ is a line not

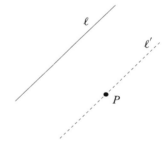

Figure 1.7. The parallel postulate.

---

[3]The word "axiom" derives from a Greek word meaning "something worthy."

[4]The word "postulate" derives from a medieval Latin word meaning "to nominate" or "to demand."

[5]John Playfair (1748–1819) was an eighteenth century geometer at the university of St. Andrews. People don't realize that Euclidean geometry as we use it today is *not* as Euclid wrote it. Playfair and Hilbert and many others worked to modernize it and make it coherent.

passing through that point, then there is a *unique* second line ℓ' passing through P that is parallel to ℓ. See Figure 1.7. The parallel postulate is part of Euclid's geometry of 2300 years ago. And people wondered for over 2000 years whether this assertion should actually be an axiom. Perhaps it could be proved from the other four axioms of geometry (see Section 2.2 for a detailed treatment of Euclid's axioms). There were mighty struggles to provide such a proof, and many famous mistakes made (see [GRE] for some of the history). But, in the 1820s, János Bolyai and Nikolai Lobachevsky showed that the parallel postulate can never be proved. The surprising reason was that there exist models for geometry in which all the other axioms of Euclid are true, yet the Parallel Postulate is false. So the parallel postulate now stands as one of the axioms of our most commonly used geometry.

Generally speaking, in any subject area of mathematics, one begins with a brief list of definitions and a brief list of axioms. Once these are in place, accepted and understood, then one can begin posing and proving theorems. A proof can take many different forms. The most traditional form of mathematical proof is a precise sequence of statements linked together by strict rules of logic. But the purpose of this book is to discuss and consider what other forms a proof might take. Today, a proof could (and often does) take the traditional form that goes back to Euclid's time. But today there are dozens of proof techniques: direct proof, induction, proof by enumeration, proof by exhaustion, proof by cases, and proof by contradiction, to name just a few. A proof could also consist of a computer calculation. Or it could consist of the construction of a physical model. Or it could consist of a computer *simulation* or *model*. Or it could consist of a computer algebra computation using `Mathematica` or `Maple` or `MATLAB`. It could also consist of an agglomeration of these various techniques.

One of the main purposes of this book is to present and examine the many forms of mathematical proof and the role that they play in modern mathematics. In spite of numerous changes and developments in the way we view the technique of proof, this fundamental methodology remains a cornerstone of the infrastructure of mathematical thought. As we have indicated, a key part of any proof—no matter what form it may take—is logic. And what is logic? This is the subject of the next section.

The philosopher Karl Popper believed [POP] that *nothing* can ever be known with absolute certainty. He instead had a concept of "truth up to falsifiability." Mathematics, in its traditional modality, rejects this point of view. Mathematical assertions that are proved according to the accepted canons of mathematical deduction are believed to be irrefutably true. And they will continue to be true. This permanent nature of mathematics makes it unique among human intellectual pursuits.

The paper [YEH] has a stimulating discussion of different modes of proof and their role in our thinking. What is a proof, why is it important, and why do we need to continue to produce proofs?

1.3 How Do Mathematicians Work?

We all have a pretty good idea how a butcher does his work, or how a physician or bricklayer does his work. In fact we have *seen* these people ply their craft. There is little mystery or doubt as to what they do.

But mathematicians are different. They may work without any witnesses, and often prefer to be in private. Many mathematicians sit quietly in their offices or homes and think. Some have favorite objects that they fondle or toss in the air. Others doodle. Some have dartboards. In fact Field Medalist Paul J. Cohen (1924–2007) used to throw darts fiercely at a dart board, and he claimed that he envisioned throwing the darts at his brother (against whom his parents pitted him in competition).

Some mathematical scientists have had rather unpredictable and delightfully surprising ways of doing their stuff. Mathematical physicists Richard Feynman (1918–1988) liked to think about physics while being at a strip club near Caltech. He went there every afternoon. When the strip club got in trouble with the law, the only one of its esteemed customers (which included doctors, lawyers, and even priests) who felt comfortable in testifying on its behalf was Richard Feynman!

Physics Nobel Laureate Steven Weinberg (1933–   ) used to formulate his theories of cosmology while watching soap operas on television. He was absolutely addicted to *As the World Turns* and some other favorites.

Although we often tell ourselves that a mathematician just sits and thinks, this is far from the whole truth. Mathematicians take walks, play pingpong, lift weights, meditate, talk to people, give lectures, and engage in discussion and debate. They show people half-proofs and half-truths in the hope of getting some help and working the result up into a full-blown theorem. They work through ideas with their students. They run seminars. They write up sets of notes. They publish research announcements. They go to conferences and kick ideas around. They listen to others lecture.[6] They read. They search the Web. They calculate and experiment. Some mathematicians will do sophisticated simulations on the computer. Others may build physical models. My teacher Fred Almgren was fond of dipping bent wires into a soap solution and examining the resulting soap bubbles. My view is that whatever works is good. It doesn't really matter how you get there, as long as you reach the goal.

1.4   The Foundations of Logic

Today mathematical logic is a subject unto itself. It is a full-blown branch of mathematics, like geometry, or differential equations, or algebra. But, for the purposes of practicing mathematicians, logic is a brief and accessible set of rules by which we live our lives.

The father of logic as we know it today was Aristotle (384–322 BCE). His *Organon* laid the foundations of what logic should be. We consider here some of Aristotle's precepts.

---

[6]I once got a great idea that helped me solve a problem of long standing by listening to a semi-incoherent lecture by a famous French mathematician. I cannot say too much of what his lecture was about, but there was one step of his proof that clinched it for me.

"IT'S AN EXCELLENT PROOF, BUT IT LACKS WARMTH AND FEELING."

Figure 1.8.

1.4.1 The Law of the Excluded Middle

One of Aristotle's rules of logic was that every sensible statement, which is clear and succinct and does not contain logical contradictions, is either true or false. There is no "middle ground" or "undecided status" for such a statement. Thus the assertion

> If there is life as we know it on Mars, then fish can fly.

is either true or false. The statement may seem frivolous. It may appear to be silly. There is no way to verify it, because we do not know (and we will not know any time soon) whether there is life as we know it on Mars. But the statement makes perfect sense. So it must be true or false. We *do know* that fish cannot fly. But we cannot determine the truth or falsity of the displayed statement because we do not know whether there is life as we know it on Mars.

You might be thinking, "Professor Krantz, that analysis is not correct. The correct truth value to assign to this sentence is 'Undecided.' We do not know about life on Mars so we cannot decide whether this statement is true. Perhaps in a couple of centuries we will have a better idea and we can assign a valid truth value to the sentence. But we cannot do so now. Thus the vote is 'undecided.' "

Interesting argumentation, but this is not the point of view that we take in mathematics. Instead, our thought is that *God* knows everything—He knows whether there is life as we know it on Mars—therefore *He* surely knows whether the statement is true or false. The fact that *we* do not know is an unfortunate artifact of our mortality. But it does not change the basic fact that *this sentence is either true or false.* Period.

It may be worth noting that there *are* versions of logic that allow for a multivalued truth function. Thus a statement is not merely assigned one of two truth values (i.e., "true" or "false"). Other truth values are allowed. For example, the statement "Barack Obama is President" is true as I write this text, but it will not be true in perpetuity. So we could have a truth value to indicate a transient truth. The book [KRA4] discusses multivalued logics. Traditional mathematics uses a logic with only two truth values: *true* and *false*. Thus traditional mathematics rejects the notion that a sensible statement can have any undecided, or any transient, truth status.

1.4.2 *Modus Ponendo Ponens* and Friends

At this point we would like to take the reader on a brief excursion into the terminology and methodology of mathematical logics as one of the ways to know how mathematicians think.

The name *modus ponendo ponens*[7] is commonly applied to the most fundamental rule of logical deduction. It says that if we know that "**A** implies **B**" and if we know "**A**" then we may conclude "**B**." This is most commonly summarized (using the standard notation of $\Rightarrow$ for "implies" and $\wedge$ for "and") as

$$[(\mathbf{A} \Rightarrow \mathbf{B}) \wedge \mathbf{A}] \Rightarrow \mathbf{B}.$$

We commonly use this mode of inference in everyday discourse. Unfortunately, we also very frequently *mis*-use it. How often have you heard someone reason as follows?

- All *rock stars* eat breakfast.
- My distinguished opponent eats breakfast.
- Therefore my distinguished opponent is a rock star.

You may laugh, but one encounters this type of thinking on news broadcasts, in the newspaper and "Facebook". As an object lesson, let us analyze this specious reasoning.

Let[8]

$$\mathbf{A}(x) \equiv x \text{ is a rock star,}$$

$$\mathbf{B}(x) \equiv x \text{ eats breakfast,}$$

$$o \equiv \text{ my distinguished opponent.}$$

Then we may diagram the logic above as

$$\mathbf{A}(x) \Rightarrow \mathbf{B}(x)$$

$$\mathbf{B}(o)$$

$$\text{therefore } \mathbf{A}(o).$$

---

[7]The translation of this Latin phrase is "mode that affirms by affirming."

[8]Here we use the convenient mathematical notation $\equiv$ to mean "is defined to be."

Figure 1.9.

You can see now that we are misusing *modus ponendo ponens*. We have **A** $\Rightarrow$ **B** and **B** and we are concluding **A**.

This is a quite common error of confusing the converse with the contrapositive. Let us discuss this fundamental issue. If **A** $\Rightarrow$ **B** is a given implication then its *converse* is the implication **B** $\Rightarrow$ **A**. Its *contrapositive* is the statement $\sim$ **B** $\Rightarrow$ $\sim$ **A**, where $\sim$ stands for "not." One sometimes encounters the word "converse" in everyday conversation but rarely the word "contrapositive." So these concepts bear some discussion.

Consider the statement

Every healthy horse has four legs.

It is convenient to first rephrase this more simply as

A healthy horse has four legs.

If we let

$$\mathbf{A}(x) \equiv x \text{ is a healthy horse,}$$
$$\mathbf{B}(x) \equiv x \text{ has four legs,}$$

then the displayed sentence says

$$\mathbf{A}(x) \Rightarrow \mathbf{B}(x).$$

The *converse* of this statement is

$$\mathbf{B}(x) \Rightarrow \mathbf{A}(x),$$

or

An object with four legs is a healthy horse.

It is not difficult to see that the converse is a distinct statement derived from the original one that each healthy horse has four legs. And, whereas the original statement is true, the converse statement is *false*. It is *not* generally the case that a thing with four legs is a healthy horse. For example, most tables have four legs; but a table is not a healthy horse. A sheep has four legs; but a sheep is not a healthy horse.

The contrapositive is a different matter. The *contrapositive* of our statement is

$$\sim \mathbf{B}(x) \Rightarrow \sim \mathbf{A}(x),$$

or

A thing that does not have four legs is not a healthy horse.

This is a different statement from the original sentence. But it is in fact *true*. If I encounter an object that does not have four legs, then I can be sure that it is *not* a healthy horse, because in fact any healthy horse has four legs. A moment's thought reveals that in fact this contrapositive statement is saying precisely the same thing as the original statement—just using slightly different language.

In fact *it is always the case that the contrapositive of an implication is logically equivalent with the original implication.* Such is not the case for the converse.

Let us return now to the discussion of whether our distinguished opponent is a rock star. We began with $\mathbf{A} \Rightarrow \mathbf{B}$ and with $\mathbf{B}$ and we concluded $\mathbf{A}$. Thus we were misreading the implication as $\mathbf{B} \Rightarrow \mathbf{A}$. In other words, we were *misinterpreting the original implication as its converse.* It would be correct to interpret the original implication as $\sim \mathbf{B} \Rightarrow \sim \mathbf{A}$, because that is the contrapositive and it is logically equivalent with the original implication. But of course $\sim \mathbf{B} \Rightarrow \sim \mathbf{A}$ and $\mathbf{B}$ taken together do not imply anything.

The rule *modus tollendo tollens*[9] is in fact nothing other than a restatement of *modus ponendo ponens*. It says that

$$\text{If } [(\mathbf{A} \Rightarrow \mathbf{B}) \text{ and } \sim \mathbf{B}], \text{ then } \sim \mathbf{A}.$$

Given the discussion we have had thus far, *modus tollendo tollens* is not difficult to understand. For $\mathbf{A} \Rightarrow \mathbf{B}$ is equivalent with its contrapositive $\sim \mathbf{B} \Rightarrow \sim \mathbf{A}$. And if we also have $\sim \mathbf{B}$ then we may conclude (by *modus ponendo ponens*!) the statement $\sim \mathbf{A}$. That is what *modus tollendo tollens* says.

It is quite common to abbreviate *modus ponendo ponens* by *modus ponens* and *modus tollendo tollens* by *modus tollens*.

---

[9]The translation of this Latin phrase is "mode that denies by denying."

Figure 1.10.

1.5 What Does a Proof Consist Of?

Most of the steps of a mathematical proof are applications of *modus ponens* or *modus tollens*. This is a slight oversimplification, since there are a great many proof techniques that have been developed over the past two centuries (see Chapter 12 for detailed discussion of some of them). These include the numerous methods cited in 1.2. They are all built on *modus ponendo ponens*.

It is really an elegant and powerful system. *Occam's razor* is a logical principle posited in the fourteenth century (by William of Occam (1288–1348)) that advocates that a proof system should have the smallest possible set of axioms and rules of inference. That way the possibility that there are internal contradictions built into the system is minimized, making it is easier to find the source of ideas. Inspired both by Euclid's *Elements* and by Occam's razor, mathematics has striven for all of modern time to keep the fundamentals of its subject as streamlined and elegant as possible. Mathematicians want the list of definitions to be as short as possible, and to have a collection of axioms or postulates as concise and elegant as possible. If one opens up a classic text on group theory—such as Marshall Hall's masterpiece [HAL], one finds that there are just three axioms on the first page. The entire 434-page book is built on just those three axioms.[10] Or instead have a look at Walter Rudin's classic *Principles of Mathematical Analysis* [RUD]. There the subject of real variables is built on just twelve axioms. Or look at a foundational book on set theory, such as Suppes [SUP] or Hrbacek and Jech [HRJ]. There we see the entire subject built on eight axioms.

---

[10]In fact there has recently been found a way to enunciate the premises of group theory using just *one* axiom, and *not* using the word "and." References for this work are [KUN], [HIN], and [MCU].

Groups

A *group* in mathematics is a collection of objects that has an operation for combining these objects. The operation should be a reasonable arithmetic-like operation that satisfies familiar properties like associativity $(a \cdot (b \cdot c) = (a \cdot b) \cdot c)$. The group has an *identity element e* $(a \cdot e = e \cdot a = a)$. Every element of a group has an *inverse element* a^{-1} $(a \cdot a^{-1} = a^{-1} \cdot a = e)$. Some groups are *commutative* $(a \cdot b = b \cdot a)$; most are not.

There are many types of groups: groups of numbers, groups of matrics, groups of operators on a Hilbert space. Group theory is one of the great unifying abstractions of modern mathematics.

1.6 The Purpose of Proof

The experimental sciences (physics, biology, chemistry, for example) tend to use laboratory experiments or tests to check and verify assertions. The benchmark in these subjects is the *reproducible experiment with control*. In their published papers, these scientists will briefly describe what they have discovered, and how they carried out the steps of the corresponding experiment. They will describe the *control*, which is the standard against which the experimental results are compared. Those scientists who are interested can, on reading the article, then turn around and replicate the experiment in their own labs. The really classic, fundamental, and important, experiments become classroom material and are reproduced by students all over the world. Most experimental science is *not* derived from fundamental principles (like axioms). The intellectual process is more empirical, and the verification procedure is correspondingly practical and direct.

Theoretical physics is a bit different. Scientists like Stephen Hawking, Edward Witten, or Roger Penrose never set foot in a laboratory. They just *think* about physics. They rely on the experimentalists to give them fodder for their ideas. The experimentalists will also help them to test their ideas. But they themselves do not engage in the benchmarking procedure.[11]

This process works reasonably well for theoretical physics, but not always. Einstein's theory of general relativity was first promulgated in 1915, and experiments of Eddington in 1919 tended to support the idea (making Einstein instantly a celebrity). But the theory did not become well established until about 1970, with the development of ideas about black holes and quasars. The theory of *strings*, which is a fairly new set of ideas that seeks to unify the general theory of relativity with quantum mechanics,[12] has been an exciting and fundamental part of physics now for over twenty years. But there is *no* experimental

---

[11]The fascinating article [JAQ] considers whether, like physicists, mathematicians should be divided into "theorists" and "experimentalists".

[12]String theory is quite "far out." This new set of ideas is supposed to describe the fundamental composition of nature in the 3-dimensional space in which we live. It provides an explanation for the provenance of gravity. But in fact a string lives in either ten- or twenty-six-dimensional space! A compelling description of string theory, and its significance for our world, may be found in [GRE].

evidence to support any of the ideas of string theory. In some sense, string theory is a set of ideas waiting to be born.

Mathematics is quite a different kind of intellectual enterprise. In mathematics we set our definitions and axioms in place *before* we do anything else. In particular, *before we endeavor to derive any results* we must engage in a certain amount of preparatory work. Then we give precise, elegant formulations of statements and we prove them. Any statement in mathematics that lacks a proof has no currency.[13] Nobody will take it as valid. And nobody will use it in his/her own work. The proof is the final test of any new idea. And, once a proof is in place, that is the end of the discussion. Nobody will ever find a counterexample, or ever gainsay that particular mathematical fact.

It should not be thought that the generation of mathematical proofs is mechanical. Far from it. A mathematician discovers ideas *intuitively*—like anyone else. He or she will just "see" or "sense" that something is true. This will be based on experience and insight developed over many years. Then the mathematician will have to think about why this new "fact" is true. At first, the sketch of the idea of a proof may be jotted down. Over a long period of time, additional ideas will be generated and pieces of the proof will be assembled. In the end, all the details will be filled in and a bona fide, rigorous proof, following the strict linear dictates of logic, will be the result.

Throughout history there have been areas of mathematics where we could not decide what a proof was. We had neither the language nor the notation nor the concepts to write *anything* down rigorously. Probability suffered many missteps, and was fraught with conundrums and paradoxes for hundreds of years until Andrei Kolmogorov (1903–1987) realized in the 1930s that measure theory was the right tool for describing probabilistic ideas. In the 1930s the Italian algebraic geometers used to decide what a "theorem" was by getting together in a group, discussing the matter, and then taking a vote. In fact it was worse than that. There was considerable, intense rivalry among Federigo Enriques (1871–1946), Guido Castelnuovo (1865–1952), and others. They would prove new results and announce them, but refuse to show anyone the proofs. So the Italian mathematicians would discuss the matter, offer speculations about how the new results could be proved, and then take a vote. One upshot of this feuding was that there was no verification process; there was no development of technique. The subject stagnated. The (currently-believed-to-be) correct tools for doing algebraic geometry were developed many years later by André Weil (1906–1998), Alexandre Grothendieck (1928–), Oscar Zariski (1899–1986), Jean-Pierre Serre (1926–), Claude Chevalley (1909–1984), and many others.

Gary Miller and Michael Rabin introduced a new twist into our subject by producing (in 1976) a *probabilistic proof* of a mathematical theorem. They studied how to prove whether a certain large number p is prime, by devising an iterative procedure with the property that each application of the algorithm increases the probability that the number is prime (or else shows that it is not). So, with enough iterations, one could make the probability as close to 1 as desired. But their method could never be used to achieve

---

[13]But mathematicians definitely engage in heuristics, algorithmic logic, and conjectures. These are discussed elsewhere in this book.

Probability Theory

Probability theory concerns itself with the likelihood that certain events will occur. If we flip a coin, then what is the likelihood that it will come up "heads"? If we draw two cards from a standard deck of 52 cards, what is the likelihood that they will form a pair? In modern physics, many quantum mechanical statements are of necessity probabilistic (this is in part what the Heisenberg uncertainty principle tells us). Gene matching—such as is used in courts of law to provide evidence against criminals—involves considerable probability theory (because the billions of gene sites can be matched only to within a measurable certitude).

Algebraic Geometry

Algebraic geometry is concerned with studying the zero sets of polynomials. If $p(x) = a_0 + a_1 x + \cdots + a_k x^k$ is a polynomial, then what can we say about the collection of numbers x such that $p(x) = 0$?

The questions become particularly sticky for polynomials of two or more variables. As an instance, the polynomial $p(x, y) = x^2 + y^2 + 1$ has no zeros if we restrict the variables x and y to be real numbers. Expanding to the complex numbers, we find that the zero set forms a surface in four-dimensional space.

mathematical certainty (unless the answer was negative). Their work strengthens earlier results of Robert Solovay and Volker Strassen.

In fact, the Miller–Rabin ideas fueled an interesting mathematical controversy that took place in the early 1970s. Rutgers mathematician Raphael Zahler published an announcement [ZAH] and then a detailed proof [THZ] in which he showed that a certain homotopy group was nonzero. At the same time, Japanese mathematicians S. Oka and H. Toda published a paper [OKT] in which they asserted that that same homotopy group *was zero*.

The mathematical experts (and many others as well) studied both papers to see where the problem was. Obviously both theorems could not be true. But the error (in one of the proofs) could not be found. Topologist J. Frank Adams, in his review of the article [ZAH], asserted that he had done the calculation independently and could confirm Zahler's result. Ultimately, Oka and Toda acknowledged their error in July of 1974 and withdrew their claim. So all is well.

But not quite. Gina Kolata wrote an article [KOL] about the controversy. Citing the work of Miller and Rabin, she suggested that if we accept probabilistic proofs, then the ideas in such work would be much more accessible and transparent, and we would not have controversies like this. The *New York Times* picked up on the issue, and published an editorial suggesting that mathematical proofs are so long and complicated that nobody can understand them anymore (this in spite of the fact that Zahler's proof was only 13 pages!).

Event	Date
Gödel publishes incompleteness theorem	1931 CE
First Bourbaki book appears	1939 CE
Erdős and Selberg give elementary proof of prime number theorem	1949 CE
von Neumann and Goldstine produce stored-program computer	1952 CE
Newell and Simon produce the Logic Theory Machine	1955 CE
Gelernter produces Geometry Machine	1959 CE
Gilmore, Wang, and Prawitz create theorem-proving machines	1960 CE
Theorem prover SAM V in use	1966 CE
Robinson introduces nonstandard analysis	1966 CE
Bishop publishes *Foundations of Constructive Analysis*	1967 CE
de Bruijn introduces proof checker `Automath`	1967 CE
Cook invents **NP**-completeness	1971 CE
Overbeek creates theorem prover Aura	1972 CE
Guilloud and Bouyer compute π to 1,000,000 digits	1973 CE
Appel and Haken give computer proof of four-color theorem	1976 CE
Jobs and Wozniak invent the mass-market personal computer	1977 CE
Thurston formulates the geometrization program	1980 CE
Knuth invents $\TeX$	1984 CE
De Branges proves Bieberbach conjecture	1984 CE
automated proof by Shankar of Gödel incompleteness theorem	1986 CE
L. Paulson and T. Nipkow introduce `Isabelle` theorem prover	1988 CE
M. Gordon introduces `HOL` theorem prover	1988 CE
Kanada and Tamura compute π to 1,000,000,000 digits	1989 CE
Hoffman, Hoffman, & Meeks use computer to generate embedded minimal surfaces	1990 CE
A. Trybulec introduces `Mizar` theorem prover	1992 CE
Horgan publishes *The Death of Proof?*	1993 CE
Hsiang publishes "solution" of Kepler problem	1993 CE
Andrew Wiles proves Fermat's last theorem	1994 CE
Finite simple groups classified	1994 CE
McCune creates `Otter` theorem-proving software	1994 CE
McCune proves Robbins conjecture using EQP	1997 CE
Almgren writes 1728-page regularity paper	1997 CE
Hsiang publishes book with "definitive" proof of Kepler	2001 CE
Kanada, Ushiro, and Kuroda compute π to 1,000,000,000,000 digits	2002 CE
Perelman announces proof of Poincaré conjecture	2004 CE
Hales and Ferguson give computer solution of Kepler's problem	2006 CE

all of basic mathematics, using only logic, from a minimal set of axioms. The point was to use strict rules of logic at every step, and virtually *no words*! Only symbols! The end result was a massive, three-volume work that is practically unreadable. This was an exercise in pure mathematical logic taken to an exquisite extreme. One of the payoffs in this book, after about 1200 pages of hard work, was the theorem

$$2 + 2 = 4.$$

In his autobiography, Russell allowed that he "never quite recovered from the strain" of writing this monumental work. After it was completed he gave up mathematics and became a full-time philosopher.

There are not many people today—even mathematicians—who study the book of Whitehead and Russell. But it is an important step in the development of mathematical rigor, and of what a proof should be. It represented, in its time, a pinnacle of the power of abstract logic. Today there is software, such as `Isabelle`, that will take a mathematical proof (expressed in the language that modern mathematicians use in a published paper) and translate it to a *formal* proof, in the vein of Whitehead and Russell or of the Zermelo–Fraenkel axioms of set theory.

It might be stressed that what Whitehead and Russell were doing was to produce a *strictly formal* development of mathematics. Their purpose was not to produce something readable, or comprehensible, or educational.[17] Their purpose was to archive mathematics for the record, using the strictest rules of formal logic. A mathematical paper written today in the style of Whitehead and Russell would not be published. No journal would consider it—because this mode of expression is not effective mathematical communication.

The mathematical proofs that we *do* publish today are distinctly less formal than the Whitehead–Russell model. Even though we adhere to strict rules of discourse, we also leave out steps and make small leaps and leave details to the reader—because we want to get the ideas across in the most concise and elegant and effective manner possible. Usually what we are publishing is a toolkit with which the reader can assemble his/her own proof. Such is quite analogous to what the chemist does: by publishing a paper that sketches how a certain experiment was performed (and describes what conclusions were drawn from it) so that the interested reader may reproduce the experiment. Often an important chemistry paper, describing years of hard work by dozens of people, will be just a few pages in length. This is an extreme application of Occam's razor: the key ideas are recorded, so that other scientists may reproduce the experiment if needed.

1.8 Platonism versus Kantianism

A question that has plagued philosophers of mathematics for many centuries, and which has heated up considerably in recent years is whether mathematics is a Platonic or a Kantian activity. What does this mean?

---

[17]In fact—and this seems astonishing considering how important this work is—they had a difficult time getting *Principia Mathematica* published. They actually had to foot part of the publication cost themselves!

The Platonic view of the world is that mathematical facts have an independent existence—very much like classical Platonic ideals—and the research mathematician *discovers* those facts—very much as Amerigo Vespucci discovered America, or Jonas Salk discovered his polio vaccine.

The Kantian view of the world is that the mathematician creates the subject from within. The idea of set, of group, of pseudoconvexity are all products of the human mind. They do not exist out there in nature. We (the mathematical community) have *created* them.

My own view is that both these paradigms are valid, and both play a role in the life of any mathematician. On a typical day some mathematicians go to their offices, sit down and think, or examine mathematical ideas that already exist and can be found in some paper written by some other mathematician. But others will also cook things up from whole cloth—maybe create a new axiom system, or define a new concept, or formulate a new hypothesis. These two activities are by no means mutually exclusive, and they both contribute to the rich broth that is mathematics.

The Kantian position raises interesting epistemological questions. Do we think of mathematics as being created from the ground up by each individual? If that is so, then there are hundreds, if not thousands, of distinct individuals creating mathematics from within. How can they communicate and share their ideas? Or perhaps the Kantian position is that mathematics is created by some shared consciousness of the aggregate humanity of mathematicians. And then is it up to each individual to "discover" what the aggregate consciousness has been creating? This is starting to sound awfully Platonic.

The Platonic view of reality seems to border on theism. For if mathematical truths have an independent existence—floating out there in the ether somewhere—then who created those truths? And by what means? Is it some higher power, with whom we would be well advised to become better acquainted? It can be argued that, once a mathematical notion or system is axiomatized, then all subsequent results *exist* Platonically; therefore mathematicians *discover* these ideas and their proofs. The Kantian view fades into the background. The art of mathematics is to determine which systems, which theorems, and which proofs are interesting.

The Platonic view makes us more like physicists. It would not make much sense for a physicist to study his subject by simply making things up—or cooking them up through pure cogitation. For the physicist is supposed to be describing the world around him. A physicist like Stephen Hawking, who is very creative and imaginative is capable of dreaming up ideas like "black hole," "supergravity," and "wormholes," but these are all intended to help explain how the universe works. They are not like manufacturing a fairy tale.

There are philosophical consequences for the thoughts expressed in the last paragraph. Physicists do not feel honor-bound to prove the claims made in their research papers. They frequently use other modes of discourse, ranging from description to analogy to experiment to calculation. If we mathematicians are Platonists, describing a world that is "already out there," then why can we not use the same discourse that the physicists use? Why do we need to be so wedded to proofs?

A very thoughtful and enlightening discussion of these issues can be found in [MAZ]. It will be some time before we know the definitive answers to the questions raised here.

1.9   The Experimental Nature of Mathematics

The discussion in the last couple of sections was accurate and fairly complete, but it was not entirely truthful. Mathematicians *do* in fact engage in experimentation. How does this fit in with the strictly rigorous, axiomatic methodology described so far?

What we have been discussing is the way that mathematics is *recorded*. It is because we use the axiomatic method and *proof* to archive our ideas that our subject is reliable, reproducible, and infallible. Mathematical ideas travel well, and stand up to the test of time, because they are recorded in a bulletproof format. But this is not the way that mathematical facts are *discovered*. The working mathematician discovers new mathematical truths by *trying things*: working examples, talking to people, making conjectures, giving lectures, attempting to formulate results, taking a stab at proofs, deriving partial results, making mistakes,[18] and trying to learn from those mistakes. Probably the first ten attempts at the formulation of a new theorem will not be quite right. Hypotheses will have to be modified, perhaps strengthened. Conclusions may have to be altered or weakened. A theorem is arrived at and captured, and finally formulated by means of trial and error. It frequently happens that the experienced mathematician will *know* something is true—will be able to picture it and describe it—but will not be able to formulate it precisely. It will be impossible at first to write down a rigorously formulated theorem.

In fact this is one of the most remarkable things about the professional mathematician. A lifetime can be spent making mistakes and trying to learn from them. There is hardly any other profession that can make this claim. The mathematician in pursuit of a particular holy grail—a new theorem, or a new theory, or a new idea—could easily spend two or three years or more experimenting and trying things that do not work, failings and starting over.

But here is the point. Once the mathematician figures out what is going on, and then finally arrives at a rigorous formulation and proof of a new idea, then it is time to engage the axiomatic method. The key idea is that the methodology of

$$\text{Definitions} \Rightarrow \text{Axioms} \Rightarrow \text{Theorem and Proof}$$

is the way that we *record* mathematics. It is the way that we can ensure permanence for our ideas, so that they will travel to and be comprehensible to future generations. It is *not* the way that we discover mathematics.

Today there is a remarkable mathematics journal called *Experimental Mathematics*. This journal—in a constructive way—flies in the face of mathematical tradition. The tradition—going back to Euclid—has been to write up mathematics for the permanent record in a rigorous, formalized, axiomatic manner. One does so in such a way as *not* to reveal anything about how the idea was arrived at, or about how many failed attempts there were, or what the partial results might have been. In short, a published mathematical paper is like a gleaming crystal ball, and the rest of the world is on the outside looking in.[19]

---

[18] An old joke has it that a mathematician is cheap to fund because all that is needed is paper, pencil, and a trash can. A philosopher is even cheaper because the trash can is not needed.

[19] There is a grand tradition in mathematics of *not* leaving a trail of corn so that the reader may determine how the mathematical material was discovered or developed. Instead, the reader is supposed to figure it all out for himself.

The periodical *Experimental Mathematics* turns the archetype just described on its head. For this forum encourages reports of partial results, descriptions of data generated by computer experiment, ideas learned from graphical images, assessments of numerical data, and analyses of physical experiments. The journal encourages speculation, and the presentation of partial or incomplete results. It publishes, for the most part, papers that other traditional mathematics journals will not consider. It has taken a bold step in acknowledging a part of the mathematical process that has never been formally accepted. And in doing so it has made a substantial and lasting contribution to our literature.

Experimental Mathematics is published by A K Peters. Klaus and Alice Peters both have mathematics pedigrees; they have shown insight into the process of disseminating mathematical ideas. This journal is one of their innovations.

1.10   The Role of Conjectures

The highest and finest form of guidance or direction that one mathematician can provide to the community of mathematicians and to students is to prove a great theorem. This gives everyone something to think about; it points to new directions of research, and it raises ever more questions that are grist for the collective mill. But there are many other ways in which a mathematician can contribute to the common weal, exert some influence over the nature of the subject, and make a difference in the directions of research. An instance of this type of activity is the formulation of conjectures.

A mathematician who has worked in a particular subject area for some years will have a very strong sense of how the ideas fit together, which concepts are important, and which questions are the guiding principles for the subject. *If* that mathematician is a recognized leader or visionary in the field, whose opinions carry some weight, *if* perhaps this person is recognized as one of the creators of the subject area, then this person has the *prerogative* to make one or more conjectures (which other mathematicians will take seriously).

A *conjecture* is the postulation that something ought to be true (or perhaps false). A common way to pose a conjecture is to write a good paper and, in summary remarks at the end, say, "Here is the direction that I think the subject ought to go now. Here is what I think is true." And then you make a formal enunciation of a result. This is a result that you *cannot prove*—although you may be able to offer a plausibility argument, or the proof of a partial result, or at least some supportive evidence. Such a conjecture can have considerable influence over a subject, and can cause a good many people to shift the direction of their research.

Even though the academic subject of mathematics has few set rules, and even though there is room to let a thousand flowers bloom, it is well understood in the discipline that persons of some substance ought to be the ones making conjectures. If everyone were running around making conjectures, then the subject at hand would turn into a chaotic maelstrom, nobody would know what was true and what was false, people would get confused, and

---

The result is a Darwinian world of survival of the fittest: only those with real mathematical talent can make their way through the rigors of the training procedure. As we note elsewhere, Gauss liked to observe that an architect does not leave up the scaffolding so we can see how his building was constructed.

little progress would be made. So there is a *sotto voce* understanding that only certain kinds of people ought to make conjectures. Mathematician Saunders Mac Lane enunciated the thought quite congently as follows:

> *Conjecture* has long been accepted and honored in mathematics, but the customs are clear. If a mathematician has really studied the subject and made advances therein, then he is entitled to formulate an insight as a conjecture, which usually has the form of a specific proposed theorem. . . . But the next step must be proof and not more speculation.
>
> —Saunders Mac Lane

Sometimes, if an eminent mathematician thinks he has proved a great result but turns out to be mistaken, then the mathematical community will exhibit its deference to the individual and call the result a conjecture named after that scholar. This is what happened with the Poincaré conjecture—discussed below. Poincaré thought he had proved it, but an error was discovered. So we now all tend to believe the result—because Henri Poincaré did. And we call it the Poincaré conjecture. Recently it seems that Grigori Perelman has finally succeeded in proving the Poincaré conjecture (see Section 10.5).

Another instance of mathematical conjecture is the Riemann hypothesis. Riemann introduced in his paper [RIE] the basic ideas connected with the Riemann zeta function. He made speculations about the location of the zeros of this function (that is what the Riemann hypothesis is about). He went on to say that it would be desirable to have proofs of these assertions; he concludes by saying that his attempts at proof were unfruitful. He states that he will put these matters aside, as they are not germane to his primary goal (which was to prove the Prime Number Theorem). Sadly, Riemann died before he could accomplish his mission.

1.10.1 Applied Mathematics

Up until about 1960, the vast majority of mathematical research in the United States was in pure mathematics. The tradition in Europe was rather different. Isaac Newton and Pierre de Fermat studied optics, Newton and Sonya Kovalevskaya studied celestial mechanics, George Green studied mathematical physics, Laplace studied celestial mechanics, Poincaré treated fluid mechanics and special relativity, Gauss contributed to the theory of geodesy and astrophysics, Turing worked on cryptography, and Cauchy even helped to develop the port facility for Napoleon's English invasion fleet. There have been several other instances throughout history of good mathematicians who were also interested in physics and engineering.

But in the early 1960s and even earlier, few mathematics departments in the United States had any faculty member who interacted with people from the physics or engineering departments. Mathematicians in those days were content to sit in their offices and prove theorems about pure mathematics. They obtained the occasional diversion from chatting with their colleagues *in the mathematics department*. But in those days collaboration was the exception rather than the rule, so for the most part the mathematician was the ascetic lone wolf.

Starting in the early 1970s there was a distinct change in the viewpoint of what modern mathematics should be. Government funding agencies began to put pressure on universities and on mathematics departments to develop "applied mathematics." Here *applied mathematics* is mathematics that is used on real-world problems. We mathematicians have long contented ourselves with saying that *all* mathematics can be applied; but this process sometimes takes a while, and it is not our job to worry about what the mathematics we are doing is good for or how long it will take for the applications to develop.[20] We took comfort in citing all the many applications of Isaac Newton's mathematics, all the good uses that are made of George Green's, William Rowan Hamilton's, and Arthur Cayley's mathematics. It was well known that the Courant Institute of Mathematical Sciences in New York City is a wellspring of excellent applied mathematics. That should be enough for the entire profession.

But no. The new ideal was that every mathematics department in the United States should have applied mathematicians. These should be people who interact with researchers in other departments, people who can teach the students how mathematics is applied to the study of nature. This new mission, backed by the power of funding (or potential loss thereof), caused quite a tumult within the professoriate. Where were we to find all these applied mathematicians? Who were they? And how does one recognize excellence in applied mathematics? What are the important problems in applied mathematics? How does one study them? And how can one tell when an applied mathematician has reached a solution?

It is safe to say that for a period of fifteen or twenty years most of the mathematics departments in this country struggled with the issues just described. The Courant Institute very definitely played a significant role in supplying the needed properly trained applied mathematicians. Great Britain, long a bastion of practical science,[21] was also a great source for applied mathematical scientists. But mathematics departments had to rethink the way that they did things. A tenure case in applied mathematics does not necessarily look like a tenure case in pure mathematics. The work is published in different journals, and new standards are applied to evaluate the work. An applied mathematician does not necessarily enunciate and prove crisp, new theorems, but instead he or she might be an expert in the analysis of numerical data, or in producing graphical images of physical phenomena. The applied mathematician might create a new high-level computer language (such as John Kemeny's participation in the creation of BASIC), or collaborate with engineers or physicists or medical researchers or workers in the school of social research.

Today, after a protracted struggle, it is safe to say that the American mathematical community has embraced applied mathematics. Some few universities have separate pure and applied mathematics departments. Nowadays pure mathematicians can remain pure and applied mathematicians can do what they want to do. But most universities have just one mathematics department, and the pure and applied mathematicians coexist. This author's university's College of Arts and Sciences has just one math department, and almost all

---

[20] In his charming autobiographical memoir [HAR], British mathematician G. H. Hardy virtually crows that he has never done anything useful and never will.

[21] Ernest Rutherford (1871–1937) was a role model for the down-to-earth British scientist. He would never accept relativity theory, for example.

of its denizens are classically trained pure mathematicians. But a significant number have developed interests in applied mathematics. Two, who were originally trained in group theory and harmonic analysis, now study statistics. They collaborate with members of the Medical School and the School of Social Work. One, originally trained in harmonic analysis, is now an expert in wavelet algorithms for image compression and signal processing, consulting with many engineering firms and with professors in engineering and the Medical School. One, (namely, this author) collaborates with plastic surgeons. Another works with chemical engineers.

This is exactly the kind of symbiosis that the government and university administrations were endeavoring to foster thirty years ago. And it has come to pass. And the good news is that we are creating new courses and new curricula to validate the change. So students today are being exposed not just to pure, traditional mathematics, but also to the manifold ways in which mathematics is used. We—the mathematics departments, the students, the government, and the university administration—can point with pride to the ways in which mathematics has affected our world:

- Mathematicians designed the carburetion system in the Volvo automobile.
- Mathematical theory underlies the design of the cellular telephone.
- Mathematics is the basis for America's preeminence in radar and scanning technologies.
- Mathematical theory underlies the technology for CD music disks and DVD movie disks.
- Mathematics is the underpinning for queuing theory, coding theory, and the ideas behind Internet routing and security.
- The entire theoretical basis for cryptography is mathematical.
- Mathematics is very much in the public eye because of the Olympiad (the international mathematics contest), because of the movies *Good Will Hunting* and *A Beautiful Mind*, because of the television show *Numbers*, because of the play *Proof.*

The list could go on and on.

It is safe to say that today pure mathematics and applied mathematics coexist in a mutually nurturing environment. They do not simply tolerate each other; in fact, they provide ideas and momentum for each other. It is a fruitful and rewarding atmosphere in which to work, and it continues to develop and grow.

It has been noted elsewhere that the tradition in mathematics has been for the mathematician to be a single-combat warrior, where in an office or at home, thinking lone thoughts, and proving theorems. In 1960, and before, almost all mathematical published work had just one author. Today that has changed. In fact, in the past fifteen years the majority of mathematical work has been done collaboratively. Now the lone worker is the exception. What is the reason for this change?

First, the symbiosis between the pure and the applied worlds has necessitated that people *talk* to each other. This author works on research projects with plastic surgeons. They do not have any expertise in mathematics and I do not have any expertise in plastic surgery. So collaboration is necessary. My colleagues who work with chemical engineers or with physicists must share skills and resources in the same way and for the same reason.

But unquestionably, even among pure mathematicians, collaboration has increased dramatically. The reason is that mathematics as a whole has become more complex. In the past forty years we have learned of a great many synergies between different mathematical fields. So it makes much more sense for a topologist to talk to an analyst, or a geometer to talk to a differential equations expert. The consequence is a blooming of joint work that has enriched our subject and dramatically increased its depth.

Mathematical collaboration has sociological and psychological consequences as well. It is difficult and depressing for some mathematicians to work in a solitary fashion. The problems are difficult and discouraging, and it is easy to feel a profound sense of isolation, and ultimately of depression and failure. A good collaborator can keep one alive, and provide momentum and encouragement. We as a profession have discovered the value—both professionally and emotionally—of having collaborators and doing joint work. It has been good for all concerned. The books [KRA2] and [KRA3] discuss the nature of the collaborative process in mathematics.

1.11 Mathematical Uncertainty

In this section we explore another aspect in which we have not been entirely truthful. While it is the case that in many respects mathematics is the most reliable, infallible, reproducible set of ideas ever devised, it also contains some pitfalls (see [KLN]). In particular, the twentieth century has given mathematics some kicks in the pants. We shall take a few moments to describe some of them.

First, some background. When we write up a proof, so that it can be submitted to a journal and refereed and ultimately (we hope) published, then we are anticipating that the (mathematical) world at large will read and appreciate it, and validate it. This is an important part of the process that is mathematics: it is the mathematics profession, taken as a whole, which decides what is correct and valid, and also what is useful, interesting, and has value. The creator of new mathematics has the responsibility of setting it before the mathematical community; but the community itself either makes the work part of the canon or it does not.

Writing a mathematical paper is to walk a fine line. The rhetoric of modern mathematics is a very strict mode of discourse. On the one hand, the formal rules of logic must be followed. The paper must contain no "leaps of faith" or guesses, or sleights of hand. On the other hand, if the writer *really* includes every step, *really* cites every rule of logic being used, *really* leaves no stone unturned, then even the simplest argument could drag on for pages. A substantial mathematical theorem could take hundreds of pages to prove. This simply will not work. Most mathematical journals will not publish so much material for the few who will be able to read it. So what we do in practice is that we skip steps. Usually these are small steps (at least small in the mind of the writer), but it is not uncommon for a mathematician to spend a couple of hours working out a step in a long paper because the author left something unsaid.[22]

---

[22] What we are describing here is not entirely different from what is done in the laboratory sciences. When a chemist performs an important experiment, and writes it up for publication, it is done in a rather telegraphic style. The idea is to give the reader enough of an idea of how the experiment was done so that it can be replicated if so desired.

To summarize: we *usually* omit many steps in proofs. In principle, the reader can fill in the missing steps. Mathematicians generally find it inappropriate to leave a trail of hints so that the reader can see how the ideas were discovered. And we believe firmly in omitting "obvious steps." We leave many details to the reader. We exhibit the finished product, gleaming and elegant. We do not necessarily tell the reader how we got to the finish line.

Let us now introduce some terminology. In mathematics, a *set* is a collection of objects. This is an example of a mathematical definition—one that describes a new concept (namely "set") in everyday language. We usually denote a set by a capital roman letter, such as S or T or U. There is a whole branch of mathematics called "set theory," and it is the very foundation of most any subject area of mathematics. Georg Cantor (1845–1918) was arguably the father of modern set theory. Thus the late nineteenth and early twentieth centuries were the heydays for laying the foundations of set theory.

This book is not the place to engage in a treatment of basic set theory. But we shall introduce one auxiliary piece of terminology that will be useful in the ensuing discussion. Let S be a set. We say that x is an *element* of S, and we write $x \in S$, if x is one of the objects in S. As an example, let S be the set of positive whole numbers. So

$$S = \{1, 2, 3, \dots\}.$$

Then 1 is an element of S. And 2 is an element of S. And 3 is an element of S, and so forth. We write $1 \in S$ and $2 \in S$ and $3 \in S$, etc. But note that π is *not* an element of S. For $\pi = 3.14159265\dots$. It is not a whole number, so it is not one of the objects in S. We write $\pi \notin S$.

Now let us return to our discussion of the saga of set theory. In 1902, G. Frege (1848–1925) was enjoying the fact that the second volume of his definitive work *The Basic Laws of Arithmetic* [FRE2] was at the printer when he received a polite and modest letter from Bertrand Russell offering the following paradox:[23]

> Let S be the collection of all sets that are not elements of themselves. Can S be an element of S?

Why is this a paradox?[24]

Here is the problem. If $S \in S$ then, by the way that we defined S, S is *not* an element of S. And if S is *not* an element of S, then, by the way that we defined S, it follows that S *is* an element of S. Thus we have a contradiction no matter what.

Here we are invoking Archimedes's law of the excluded middle. It *must* be the case that either $S \in S$ or $S \notin S$, but in fact either situation leads to a contradiction. And that is

---

[23]This paradox was quite a shock to Frege.

[24]The thoughtful reader may well wonder whether it is actually possible for a set to be an element of itself. This sounds like a form of mental contortionism that is implausible at best. But consider the set S described by

The collection of all sets that can be described in fewer than fifty words.

Notice that S is a set, and S can be described in fewer than fifty words. Aha! So S is certainly an element of itself.

Russell's paradox. Frege had to rethink his book, and make notable revisions, in order to address the issues raised by Russell's paradox.[25]

After considerable correspondence with Russell, Frege modified one of his axioms and added an appendix explaining how the modification addresses Russell's concerns. Unfortunately, this modification nullified several of the results in Frege's already-published Volume 1. Frege's second volume *did* ultimately appear (see [FRE2]). Frege was somewhat disheartened by his experience, and his research productivity went into a marked decline. His planned third volume never appeared.

Lésniewski proved, after Frege's death, that the resulting axiom system that appears in print is inconsistent. Frege is nonetheless remembered as one of the most important figures in the foundations of mathematics. He was one of the first to formalize the rules by which mathematicians operate, and in that sense he was a true pioneer. Many scholars hold that his earlier work, *Begriffsschrift und andere Aufsätze* [FRE1], is the most important single work ever written in logic. It lays the very foundations of modern logic. Paul J. Cohen, one of the most distinguished logicians of the twentieth century, describes Frege's contribution in this way:

> With the publication of Frege's epic work *Begriffsschrift* in 1879, the notion of a formal system was given a definitive form. Important related work was done by Boole, and Peirce, and later Peano presented a similar approach, but with Frege's work, for the first time in the history of human thought, the notion of logical deduction was given a completely precise formulation. Frege's work not only included a description of the language (which we might nowadays call the "machine language"), but also a description of the rules for manipulating this language, which is nowadays known as predicate calculus. ... But this was a major landmark. For the first time one could speak precisely about proofs and axiomatic systems. The work was largely duplicated by others, e.g., Russell and Whitehead, who gave their own formulations and notations, and even Hilbert made several attempts to reformulate the basic notion of a formal system.

A more recent 1995 paper of G. Boolos [BOO] makes considerable strides in rescuing much of Frege's original program that is represented in the two-volume work [FRE2].

Now that we have had nearly a century to think about Russell's paradox, we realize that what it teaches us is that we cannot allow sets that are *too large*. The set S described in Russell's paradox is unallowably large. In a rigorous development of set theory, there are very specific rules for which sets we are allowed to consider and which not. In particular, modern set theory does not allow us to consider a set that is an element of itself. We will not indulge in the details here.

It turns out that Russell's paradox is only the tip of the iceberg. Nobody anticipated what Kurt Gödel (1906–1978) would teach us thirty years later.

---

[25] A popular version of Russell's paradox goes like this. A barber (assumed to be a male) in a certain town agrees to shave every man in the town who does not shave himself. He will not touch any man who ever deigns to shave himself. Who shaves the barber? If the barber shaves himself, then he himself is someone whom the barber has agreed not to shave. If instead the barber does not shave himself, then he himself is someone whom the barber must shave. A contradiction either way. So no such barber can exist.

Set Theory

Set theory is about collections of objects. Such a collection is called a *set*, and each of the objects in it is called an *element* of that set. Obviously mathematics deals with sets of various sizes and shapes. A set could be a set of points or a set of numbers or a set of triangles or a set of many other things. Of particular interest are *very large* sets—sets with infinitely many elements.

In informal language, what Gödel showed us is that—in any sufficiently complex logical system (i.e., at least as complex as arithmetic)—there will be a sensible, true statement that we cannot prove within that system.[26] This is Gödel's incompleteness theorem. It came as an unanticipated bombshell, and has completely altered the way that we think of our subject.[27] It should be stressed that the statement that Gödel found is not *completely unprovable*. If one transcends the specified logical system, and works instead in a larger and more powerful system, then one *can* create a proof of the Gödel statement.

In fact, what Gödel did was extraordinarily powerful and elegant. He found a way to assign a natural number (i.e., a positive whole number) to each statement in the given logical system. This number has come to be known as the *Gödel number*. It turns out then that statements in the logical system about the natural numbers are actually statements about the statements themselves. Thus Gödel is able to formulate a statement U within the logical system—*and this is simply a statement about the natural numbers*—that says, in effect, "U asserts that U itself is unprovable." This is a problem. Because if U is false then U is provable (which cannot be, because then U would be true). And if U is true then U is unprovable. So we have a true statement that cannot be proved *within the system*. The books [SMU1], [SMU2] provide entertaining and accessible discussions of Gödel's ideas.

Just as quantum mechanics taught us that nature is not completely deterministic—we cannot know everything about a given physical system, even if we have a complete list of all the initial conditions—so it is the case that Gödel has taught us that there will always be statements in mathematics that are "unprovable" or "undecidable."

It is safe to say that Gödel's ideas shook the very foundations of mathematics. They have profound implications for the analytical basis for our subject, and for what we can expect from it. There are also serious consequences for theoretical computer science—just because the computer scientist wants to know where any given programming language (which is definitely a system of reason) will lead and what it can produce.godelstheorem@Gödel's theorem!and computer science

The good news is that the consequences of Gödel's incompleteness theorem rarely arise in everyday mathematics. The "Gödel statement" is more combinatorial than analytical.

---

[26] Gödel even went so far as to show that the consistency of arithmetic itself is not provable. Arithmetic is the most fundamental and widely accepted part of basic mathematics. The notion that we can never be certain that it will not lead to a contradiction is definitely unsettling.

[27] The Gödel statement is in fact created *from the logical system itself* within which we are working.

One does not encounter such a sentence in calculus and analysis. It can, however, arise sometimes in algebra, number theory, and discrete mathematics.[28] There have been highly desirable and much-sought-after results in number theory that have been proved to be undecidable (the resolution of Hilbert's tenth problem is a striking example). And of course Gödel's ideas play a major role in mathematical logic.

1.12   The Publication and Dissemination of Mathematics

Five hundred years ago scientists were often rather secretive. They tended to keep their results and their scientific discoveries to themselves. Even when asked by another scientist for a specific piece of data, or queried about a specific idea, these scholars were often evasive. Why would serious scientific researchers behave in such a fashion?

We must understand that the world was different in those days. There were very few academic positions. Many of the great scientists did their research as a personal hobby. Or, if they were lucky, they could locate a wealthy patron who would subsidize their work. But one can see that various resentments and jealousies could easily develop. There was no National Science Foundation and no National Institutes of Health at the time of Johannes Kepler (1571–1630). Many a noted scientist would spend years struggling to find an academic position. The successful landing of a professorship involved various patronage issues and a variety of academic and nonacademic politics. Even Riemann landed a chair at Göttingen only shortly before his death.

One of the most famous scientists who was secretive about his work was Isaac Newton. Newton has been regarded as the greatest scientist who ever lived, and he produced myriad ideas that revolutionized scientific thought. He was an irascible, moody, temperamental individual with few friends. On one occasion a paper that he had submitted for publication was subjected to criticisms by Robert Hooke (1635–1703), and Newton took this badly. He published nothing for quite a long time after that. Newton's reluctance to publish meant that many of the key ideas of the calculus were kept under wraps. Meanwhile, Gottfried Wilhelm von Leibniz was independently developing the ideas of calculus in his own language. Leibniz did not suffer from any particular reticence to publish. And the publication of Leibniz's ideas caused considerable consternation among Newton and his adherents. They felt that Leibniz was attempting to abscond with ideas that were first created by Newton. One could argue in the other direction that Newton should have published his ideas in a timely manner. That would have removed all doubt about who first discovered calculus.

In the mid-seventeenth century, Henry Oldenburg was active in various scientific societies. Because of his personality and his connections Oldenburg became something of a go-between among scientists of the day. If he knew that scientist *A* needed some ideas of scientist *B*, then he would arrange to approach *B* to ask him to share his ideas. Usually Oldenburg could arrange to offer a quid for this largesse. In those days books were rare and expensive, and Oldenburg could sometimes arrange to offer a scientific book in exchange for some ideas.

---

[28]One must note, however, that the last decade has seen a great cross-fertilization of analysis with combinatorics. The statements we are making now may soon be out of date.

After some years of these activities, Oldenburg and his politicking became something of an institution. This led him to create the first peer-reviewed scientific journal in 1665. He was the founding editor of *The Philosophical Transactions of the Royal Society of London*. At the time it was a daring but much-needed invention that supplanted a semi-secret informal method of scientific communication that was both counterproductive and unreliable. Today journals are part of the fabric of our professional life. Most scientific research is published in journals of one kind or another.

In modern times, journals are the means of our professional survival. Any scientist who wants to establish a reputation must publish in scientific journals. A substantial scholarly gestalt must be established to get tenure. This means that the individual will have created some substantial new ideas in his/her field, and will have lectured about them, and published some write-up of the development of this thought. This dynamic circle of consideration came to be popularly known as popular "publish or perish." The notion is perhaps revealing of the academic minds.

For the past many years, and especially since the advent of NSF (National Science Foundation) grants, we have been living under the specter of "publish or perish." The meaning of this aphorism is that, if you are an academic, and if you want to get tenured or promoted, or you want to get a grant, or you want an invitation to a conference, or you want a raise, or you want the respect and admiration of your colleagues, then you must publish original work in recognized, refereed journals or books. Otherwise, you're outta here. Who coined the phrase "publish or perish"?

One might think that it was a President of Harvard. Or perhaps a high-ranking officer at the NSF. Or some Dean at Caltech. One self-proclaimed expert on quotations suggested to me that it was Benjamin Franklin! But, no, it was sociologist Logan Wilson in his 1942 book *The Academic Man, A Study in the Sociology of a Profession* [WILS]. He wrote, "The prevailing pragmatism forced upon the academic group is that one must write something and get it into print. Situational imperatives dictate a 'publish or perish' credo within the ranks."

Wilson was President of the University of Texas and (earlier) a student at Harvard of the distinguished sociologist Robert K. Merton. So he no doubt knew whereof he spoke.

Marshall McLuhan has sometimes been credited with the phrase "publish or perish," and it is arguable that it was he who popularized it. In a June 22, 1951 letter to Ezra Pound he wrote (using Pound's favorite moniker "beaneries" to refer to the universities):

> The beaneries are on their knees to these gents (foundation administrators). They regard them as Santa Claus. They will do "research on anything" that Santa Claus approves. They will think his thoughts as long as he will pay the bill for getting them before the public signed by the profesorry-rat. "Publish or perish" is the beanery motto.

1.13 Closing Thoughts

The purpose of this chapter has been to bring the reader up to speed on the dynamics of the mathematics progression and, to acquaint him or her with the basic tenets of mathematical thinking. The remainder of the book is an outline of the history of mathematics and an *analysis* of that mathematical thinking. Just what does a mathematician do? What is he or she trying to achieve? And what is the method for doing so?

An important epistemological point is this: Before you can convince anyone else that your theorem is true (by way of a proof, or something like a proof), you must first convince yourself. How do you do that? In the first pass, we usually do this with intuition. An experienced mathematician will just *know* that something is true because of having thought about it for a long time and can just *see* that it is correct. The next step is to be able to write something down. Here, by "something," a mathematicians means a sequence of steps that strongly resembles a rigorous proof. Having done so, the working mathematician's confidence is ratcheted up a notch. He or she may decide to give some talks about the work.

But the final arbiter of what is true and what is correct is writing down a rigorous argument that follows the strict rules of mathematical discourse. This can often take many months of protracted labor (even though the original insight only took a few days or a few weeks). That is the nature of the beast. Mathematics is an exacting taskmaster, and takes no prisoners. It is the job of the mathematician to record a proof so that others can read it and check it and confirm it.

For the theoretical mathematician, the accepted methodology is *proof*. The mathematician discovers new ideas or theories, finds a way to formulate them, and then must verify them. The vehicle for doing so is the classical notion of proof. In the ensuing chapters we explore how the concept of proof came about, how it developed, how it became the established methodology in the subject, and how it has been developing and changing over the years.

2

The Ancients

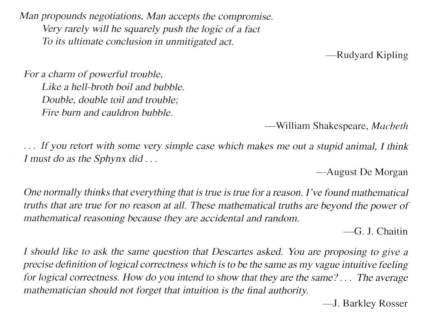

Man propounds negotiations, Man accepts the compromise.
  Very rarely will he squarely push the logic of a fact
  To its ultimate conclusion in unmitigated act.

—Rudyard Kipling

For a charm of powerful trouble,
  Like a hell-broth boil and bubble.
  Double, double toil and trouble;
  Fire burn and cauldron bubble.

—William Shakespeare, *Macbeth*

. . . If you retort with some very simple case which makes me out a stupid animal, I think I must do as the Sphynx did . . .

—August De Morgan

One normally thinks that everything that is true is true for a reason. I've found mathematical truths that are true for no reason at all. These mathematical truths are beyond the power of mathematical reasoning because they are accidental and random.

—G. J. Chaitin

I should like to ask the same question that Descartes asked. You are proposing to give a precise definition of logical correctness which is to be the same as my vague intuitive feeling for logical correctness. How do you intend to show that they are the same? . . . The average mathematician should not forget that intuition is the final authority.

—J. Barkley Rosser

2.1 Eudoxus and the Concept of Theorem

Perhaps the first mathematical proof in recorded history is due to the Babylonians. They seem (along with the Chinese) to have been aware of the Pythagorean theorem (discussed in detail below) well before Pythagoras.[29] The Babylonians had certain diagrams that indicate why the Pythagorean theorem is true, and tablets have been found to validate this fact.[30] They also had methods for calculating Pythagorean triples—that is, triples of integers (or whole numbers) a, b, c that satisfy

$$a^2 + b^2 = c^2$$

as in the Pythagorean theorem.

The Babylonians were remarkably sophisticated in a number of ways. As early as 1200 BCE they had calculated $\sqrt{2}$ to (what we would call) six decimal places.[31] They did not "prove" theorems as we conceive of the activity today, but they had well-developed ideas about mathematics (not just arithmetic).

It was the Greek Eudoxus (408–355 BCE) who began the grand tradition of organizing mathematics into theorems.[32] Eudoxus was one of the first to use this word in the context of mathematics.

Figure 2.1. The Plimpton 322 tablet.

What Eudoxus gained in the rigor and precision of his mathematical formulations, he lost because he did not prove anything. Formal proof was not yet the tradition in mathematics. As we have noted elsewhere, mathematics in its early days was a largely heuristic and empirical subject. It had never occurred to anyone that there was any need to prove anything. When you asked yourself whether a certain table would fit in your dining room, you did not prove a theorem; you just checked it out.[33] When you wondered whether a certain amount of fence would surround your pasture, you did not seek a rigorous argument; you simply unrolled the fence and determined whether it did the job. In its earliest days, mathematics was intimately bound up with questions precisely like these. Thus mathematical thinking was almost inextricable from practical thinking. And that is how its adherents viewed mathematical facts. They were just practical information, and their assimilation and verification was a strictly pragmatic affair.

---

[29] Although it must be stressed that they did not have Pythagoras's sense of the structure of mathematics, of the importance of rigor, or of the nature of formal proof.

[30] This is the famous Plimpton 322 tablet. See Figure 2.1.

[31] This can be found in the YBC 7289 tablet. Of course the Babylonians did *not* have decimal notation.

[32] The word "theorem" comes from the Greek root *theorema*, meaning "speculation."

[33] William (Willy) Feller (1906–1970) was a prominent mathematician at Princeton University. He was one of the fathers of modern probability theory. Feller and his wife were once trying to move a large circular table from their living room into the dining room. They pushed and pulled and rotated and maneuvered, but try as they might they could not get the table through the door. It seemed to be inextricably stuck. Frustrated and tired, Feller sat down with a pencil and paper and devised a mathematical model of the situation. After several minutes he was able to *prove* that what they were trying to do was impossible. While Willy was engaged in these machinations, his wife had continued struggling with the table, and she managed to get it into the dining room.

2.2 **Euclid the Geometer**

Euclid (325–265 BCE) is hailed as the first scholar to systematically organize mathematics (i.e., a substantial portion of the mathematics that had come before him), formulate definitions and axioms, and prove theorems. This was a monumental achievement, and a highly original one.

Euclid is not known as much (as were Archimedes and Pythagoras) for his original and profound insights—although there are some important theorems and ideas named after him—but he has had overall an incisive effect on human thought. After all, Euclid wrote a treatise (consisting of 13 Books)—now known as Euclid's *Elements*—which has been continuously available for over 2000 years and has been through a large number of editions. It is still studied in detail today, and continues to have a substantial influence over the way we think about mathematics.

Not a great deal is known about Euclid's life, although it is fairly certain that he had a school in Alexandria. In fact "Euclid" was a common name in his day, and various accounts of Euclid the mathematician's life confuse him with other Euclids (one a prominent philosopher). One appreciation of Euclid comes from Proclus, one of the last of the ancient Greek philosophers:

> Not much younger than these [pupils of Plato] is Euclid, who put together the *Elements*, arranging in order many of Eudoxus's theorems, perfecting many of Theaetetus's, and also bringing to irrefutable demonstration the things which had been only loosely proved by his predecessors. This man lived in the time of the first Ptolemy; for Archimedes, who followed closely upon the first Ptolemy makes mention of Euclid, and further they say that Ptolemy once asked him if there were a shorter way to study geometry than the *Elements*, to which he replied that "there is no royal road to geometry." He is therefore younger than Plato's circle, but older than Eratosthenes and Archimedes; for these were contemporaries, as Eratosthenes somewhere says. In his aim he was a Platonist, being in sympathy with this philosophy, whence he made the end of the whole *Elements* the construction of the so-called Platonic figures.

As often happens with scientists and artists and scholars of immense accomplishment, there is disagreement, and some debate, over exactly who or what Euclid actually was. The three schools of thought are these:

- Euclid was a historical character—a single individual—who in fact wrote the *Elements* and the other scholarly works that are commonly attributed to him.

- Euclid was the leader of a team of mathematicians working in Alexandria. They all contributed to the creation of the complete works that we now attribute to Euclid. They even continued to write and disseminate books under Euclid's name after his death.

- Euclid was not a historical character at all. In fact "Euclid" was a *nom de plume* adopted by a group of mathematicians working in Alexandria. They took their inspiration from Euclid of Megara (who *was* in fact an historical figure), a prominent philosopher who lived about one hundred years before Euclid the mathematician is thought to have lived.

Most scholars today subscribe to the first theory—that Euclid was a unique person who created the *Elements*. But we acknowledge that there is evidence for the other two scenarios. Almost surely Euclid had a vigorous school of mathematics in Alexandria, and there is little doubt that his students participated in his projects.

It is thought that Euclid must have studied in Plato's (430–349 BCE) academy in Athens, for it is unlikely that there would have been another place where he could have learned the geometry of Eudoxus and Theaetetus on which the *Elements* is based.

Another famous story[34] and quotation about Euclid is this: A certain pupil of Euclid, at his school in Alexandria, came to Euclid after learning just the first proposition in the geometry of the *Elements*. He wanted to know what he would gain by arduous study, doing all the necessary work, and learning the theorems of geometry. At this, Euclid called over his slave and said, "Give him three drachmas since he must needs make gain by what he learns."

What is important about Euclid's *Elements* is the paradigm it provides for the way that mathematics should be studied and recorded. He begins with several definitions of terminology and ideas for geometry, and then he records five important postulates (or axioms) of geometry. A version of these postulates is as follows:

P1   Through any pair of distinct points there passes a line.

P2   For each segment $\overline{AB}$ and each segment $\overline{CD}$ there is a unique point E (on the line determined by A and B) such that B is between A and E and the segment $\overline{CD}$ is congruent to $\overline{BE}$ (Figure 2.2).

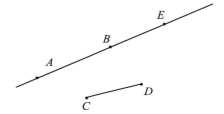

Figure 2.2.   Euclid's second postulate.

P3   For each point C and each point A distinct from C there exists a circle with center C and radius CA (Figure 2.3a).

P4   All right angles are congruent (Figure 2.3b).

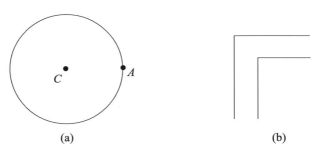

(a)   (b)

Figure 2.3.   The circle and the right angle.

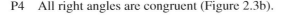

[34] A similar story is told of Plato.

These are the standard four axioms that give our Euclidean conception of geometry. The fifth axiom, a topic of intense study for 2000 years, is the so-called parallel postulate (in *Playfair's* formulation):

P5   For each line ℓ and each point P that does not lie on ℓ there is a unique line ℓ' through P such that ℓ' is parallel to ℓ (Figure 2.4).

Prior to this enunciation of his celebrated five axioms, Euclid had defined "point," "line," "circle," and the other terms that he uses. Although Euclid borrowed freely from mathematicians both earlier and contemporaneous with himself, it is generally believed that the famous "parallel postulate," that is, Postulate P5, is Euclid's own creation.

We should note, however, that the work *Elements* is not simply about plane geometry. In fact, Books VII–IX deal with number theory, in which Euclid proves his famous result that there are infinitely many primes (treated elsewhere in this text) and also his celebrated "Euclidean algorithm" for division with

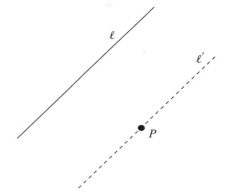

Figure 2.4.   The parallel postulate.

remainder. Book X deals with irrational numbers, and Books XI–XIII treat 3-dimensional geometry. In short, Euclid's *Elements* is an exhaustive treatment of a good deal of the mathematics that was known at the time. It is presented in a strictly rigorous and axiomatic manner that has set the tone for the way that mathematics is recorded and studied today. Euclid's *Elements* is perhaps most notable for the clarity with which theorems are formulated and proved. The standard of rigor that Euclid set was to be a model for the inventors of calculus nearly 2000 years later.

Noted algebraist B. L. van der Waerden (1903–1996) assesses the impact of Euclid's *Elements* in this way:

> Almost from the time of its writing and lasting almost to the present, the *Elements* has exerted a continuous and major influence on human affairs. It was the primary source of geometric inference, theorems, and methods at least until the advent of non-Euclidean geometry in the nineteenth century. It is sometimes said that, next to the Bible, the *Elements* may be the most translated, published, and studied of all the books produced in the Western world.

Indeed, there have been more than 1000 editions of Euclid's *Elements*. It is arguable that Euclid was and still is the most important and most influential mathematics teacher of all time. It may be added that a number of other books by Euclid survive. These include *Data* (which studies geometric properties of figures), *On Divisions* (which studies the division of geometric regions into subregions having areas of a given ratio), *Optics* (which is the first Greek work on perspective), and *Phaenomena* (which is an elementary introduction to

███

Number Theory

Number theory is concerned with properties of the whole numbers $1, 2, 3, \ldots$. Are there infinitely many prime numbers? How are they distributed? Can every even number greater than 4 be written as the sum of two odd primes? If N is a large, positive number, then how many primes are less than or equal to N?

These are great, classical questions—some of them answered today and some not. Today number theory plays a vital role in cryptography. The National Security Agency in Washington, D.C. employs over 2,000 Ph.D. mathematicians to study this subject area.

███

mathematical astronomy). Several other books of Euclid—including *Surface Loci*, *Porisms*, *Conics*, *Book of Fallacies*, and *Elements of Music*—have been lost.

2.2.1 Euclid the Number Theorist

Most of us remember Euclid's *Elements* as a work on geometry. But Books VII–IX of the *Elements* deal with number theory. One of the particular results presented there has stood the test of time, and the proof is taught today to every mathematics student. We shall discuss it now.

Recall that a *prime number* is a positive whole number that has no divisors except for 1 and itself. By tradition we do not consider 1 to be a prime. So the prime numbers are

$$2, 3, 5, 7, 11, 13, 17, 19, 23, 29, 31, 37, \ldots.$$

A number greater than 1 that is not prime is called *composite*. For example, 126 is composite. Notice that

$$126 = 2 \cdot 3^2 \cdot 7.$$

It is *not* prime. Any composite number can be factored in a unique fashion into prime factors—that is the fundamental theorem of arithmetic.

The question that Euclid considered (and, unlike many of the other results in the *Elements*, this result seems to have originated with Euclid himself) is whether there are infinitely many prime numbers. And Euclid's dramatic answer is yes.

Theorem: *There are infinitely many prime integers.*

For the proof, assume the contrary. So there are only finitely many primes. Call them $p_1, p_2, \ldots, p_N$. Now consider the number $P = (p_1 \cdot p_2 \cdots p_N) + 1$. What kind of number is P? Notice that if we divide P by p_1, then we get a remainder of 1 (since p_1 goes evenly into $p_1 \cdot p_2 \cdots p_N$). Also if we divide p_2 into P, then we get a remainder of 1. And it is the same if we divide any of p_3 through p_N into P.

Now, if P were a composite number, then it would have to be evenly divisible by some prime. But we have just shown that it is not: We have divided every known prime number into P and obtained a nonzero remainder in each instance. The only possible conclusion is that P is another prime, obviously greater than any of the primes on the original list. That is a contradiction. So there cannot be only finitely many primes. There must be infinitely many.

Euclid's argument is one of the first known instances of proof by contradiction.[35] This important method of formal proof has actually been quite controversial over the years. We shall discuss it in considerable detail as the book develops.

2.3 Pythagoras

Pythagoras (ca. 569–500 BCE) was both a person and a society (i.e., the *Pythagoreans*). He was also a political figure and a mystic. He was special in his time, among other reasons, because he involved women as equals in his activities. One critic characterized the man as "one tenth of him genius, nine-tenths sheer fudge." Pythagoras died, according to legend, in the flames of his own schools fired by political and religious bigots who stirred up the masses to protest against the enlightenment that Pythagoras sought to bring them.

The Pythagorean society was intensely mathematical in nature, but it was also quasi-religious. Among its tenets (according to [RUS]) were:

- To abstain from beans.
- Not to pick up what has fallen.
- Not to touch a white cock.
- Not to break bread.
- Not to step over a crossbar.
- Not to stir the fire with iron.
- Not to eat from a whole loaf.
- Not to pluck a garland.
- Not to sit on a quart measure.
- Not to eat the heart.
- Not to walk on highways.
- Not to let swallows share one's roof.
- When the pot is taken off the fire, not to leave the mark of it in the ashes, but to stir them together.
- Not to look in a mirror beside a light.
- When you rise from the bedclothes, roll them together and smooth out the impress of the body.

The Pythagoreans embodied a passionate spirit that is remarkable to our eyes:

Bless us, divine Number, thou who generatest gods and men.

and

Number rules the universe.

---

[35]There are many other ways to prove Euclid's result, including direct proofs and proofs by induction. In other words, it is not *necessary* to use proof by contradiction.

The Pythagoreans are remembered for two monumental contributions to mathematics. The first was establishing the importance of, and the necessity for, *proofs* in mathematics: that mathematical statements, especially geometric statements, must be verified by way of rigorous proof. Prior to Pythagoras, the ideas of geometry were generally rules of thumb that were derived empirically, merely from observation and (occasionally) measurement. Pythagoras also introduced the idea that a great body of mathematics (such as geometry) could be derived from a small number of postulates. Clearly Euclid was infuenced by Pythagoras.

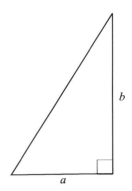

Figure 2.5.   The fraction $\frac{b}{a}$.

The second great contribution was the discovery of, and proof of, the fact that not all numbers are commensurate. More precisely, the Greeks prior to Pythagoras believed with a profound and deeply held passion that everything was built on the whole numbers. Fractions arise in a concrete manner: as ratios of the sides of triangles with integer length (and are thus *commensurable*—this antiquated terminology has today been replaced by the word "rational")—see Figure 2.5.

Pythagoras proved the result now called the *Pythagorean theorem*. It says that the legs a, b and hypotenuse c of a right triangle (Figure 2.6) are related by the formula

$$a^2 + b^2 = c^2. \qquad (*)$$

This theorem has perhaps more proofs than any other result in mathematics—well over 50 altogether. And in fact it is one of the most ancient mathematical results. There is evidence that the Babylonians and the Chinese knew this theorem at least five hundred years before Pythagoras.

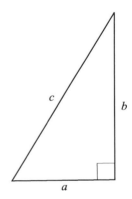

Figure 2.6.   The Pythagorean theorem.

Remarkably, one proof of the Pythagorean theorem was devised by U.S. President James Garfield (1831–1881). We now provide one of the simplest and most classical arguments.

Proof of the Pythagorean Theorem.   Refer to Figure 2.7. Observe that we have four right triangles and a square packed into a large square. Each of the triangles has legs a and b and hypotenuse c, just as in the Pythagorean theorem. Of course, on the one hand, the area of the large square is c^2. On the other hand, the area of the large square is the sum of the areas of its component pieces.

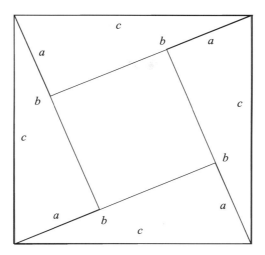

Figure 2.7. Proof of the Pythagorean theorem.

Thus we calculate that

$$c^2 = \text{(area of large square)}$$
$$= \text{(area of triangle)} + \text{(area of triangle)}$$
$$+ \text{(area of triangle)} + \text{(area of triangle)}$$
$$+ \text{(area of small square)}$$
$$= \frac{1}{2} \cdot ab + \frac{1}{2} \cdot ab + \frac{1}{2} \cdot ab + \frac{1}{2} \cdot ab + (b - a)^2$$
$$= 2ab + [a^2 - 2ab + b^2]$$
$$= a^2 + b^2.$$

This proves the Pythagorean theorem. □

It is amusing to note that (according to legend) the Egyptians had, as one of their standard tools, a rope with twelve equally spaced knots. They used this rope to form a triangle with sides 3, 4, 5—see Figure 2.8. In this way they took advantage of the Pythagorean theorem to construct right angles.

Pythagoras noticed that, if $a = 1$ and $b = 1$, then $c^2 = 2$. He wondered whether there was a rational number c that satisfied this last identity. His stunning conclusion was this:

Theorem. *There is no rational number c such that $c^2 = 2$.*

Pythagoras is telling us that a right triangle with each leg having length 1 will have a hypotenuse with irrational length. This is a profound and disturbing assertion!

Proof of the Theorem. Suppose that the conclusion is false. Then there *is* a rational number $c = \alpha/\beta$, expressed in lowest terms (i.e., α and β are integers with no factors in common) such that $c^2 = 2$. This translates to

$$\frac{\alpha^2}{\beta^2} = 2$$

or

$$\alpha^2 = 2\beta^2.$$

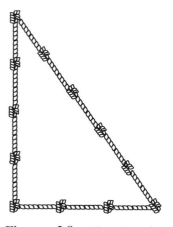

We conclude that the right-hand side is even; hence so is the left-hand side. Therefore α is even, so $\alpha = 2m$ for some integer m (see Propositions 12.1.3 and 12.1.4 below).

Figure 2.8. The Egyptian Pythagorean rope.

But then

$$(2m)^2 = 2\beta^2$$

or

$$2m^2 = \beta^2.$$

We see that the left-hand side is even, so β^2 is even. Hence β is even.

But now both α and β are even—the two numbers have a common factor of 2. This statement contradicts the hypothesis that α and β have no common factors, so it cannot be that c is a rational number; c must be irrational.   □

It is notable that T. Apostol has found a strictly graphical proof of the irrationality of $\sqrt{2}$—see [BAB1, p. 73].

The Pythagoreans realized the profundity and potential social importance of this discovery. It was ingrained in the ancient Greek consciousness that all numbers were rational. To claim the contrary would have been virtually heretical. For a time the Pythagoreans kept this new fact a secret. Ultimately, so legend has it, the Pythagoreans were destroyed by (ignorant) marauding hordes. But their universal ideas live on.

3

The Middle Ages and An Emphasis on Calculation

In any particular theory there is only as much real science as there is mathematics.

—Immanuel Kant

For, compared with the immense expanse of modern mathematics, what would the wretched remnants mean, the few isolated results incomplete and unrelated, that the intuitionists have obtained?

—David Hilbert

The sequence for the understanding of mathematics may be:

intuition, trial, error, speculation, conjecture, proof.

The mixture and the sequence of these events differ widely in different domains, but there is general agreement that the end product is rigorous proof—which we know and can recognize, without the formal advice of the logicians.

—Saunders Mac Lane

Indeed every mathematician knows that a proof has not been "understood" if one has done nothing more than verify step by step the correctness of the deductions of which it is composed and has not tried to gain a clear insight into the ideas which have led to the construction of this particular chain of deductions in preference to every other one.

—Nicolas Bourbaki

3.1 The Islamic Influence on Mathematics

The history of the Near East and the Muslims is complex, and there are many gaps. The Islamic religion was born in Arabia, and lives on there today. What some people call "Arabic mathematics" others would call "Islamic" or "Muslim" mathematics. One of the principal figures in our discussion here is Muhammad ibn Musa al-Khwarizmi. As far as we know, he was born in Baghdad around 780 CE. Thus it is not out of place to call him an Arab; but he is more commonly referred to as an "Islamic mathematician," or a "Persian mathematician." It appears that the Muslim culture was a driving force in many of the developments of the Middle Ages. So we shall adopt the policy of referring to "Islamic mathematics."

In the early seventh century, the Muslims formed a small and persecuted sect. But by the end of that century, under the inspiration of Muhammad's leadership, they had conquered lands from India to Spain—including parts of North Africa and southern Italy. It is believed that when Arab conquerors settled in new towns they would contract diseases that had been unknown to them in desert life. In those days the study of medicine was confined mainly to Greeks and Jews. Encouraged by the caliphs (the leaders of the Islamic world), these doctors settled in Baghdad, Damascus, and other cities. So we see that a purely social situation led to the contact between two different cultures, which ultimately led to the transmission of mathematical knowledge.

Around the year 800 CE, the caliph Harun al-Raschid ordered many of the works of Hippocrates, Aristotle, and Galen to be translated into Arabic. Much later, in the twelfth century, these Arabic translations were further translated into Latin so as to make them accessible to the Europeans. Today we credit the Muslims with preserving the grand Greek tradition in mathematics and science. Without their efforts much of this classical work would have been lost.

3.2 The Development of Algebra

3.2.1 Al-Khwarizmi and the Basics of Algebra

There is general agreement that the rudiments of algebra had their genesis with the Hindus. Particularly Arya-Bhata in the fifth century and Brahmagupta in the sixth and seventh centuries played a major role in the development of these ideas. Notable among the developments due to these men is the summation of the first N positive integers, and also the sum of their squares and their cubes (see our discussion of these matters in Chapter 12).

But the Islamic expansion two hundred years later caused the transfer of these ideas to the Islamic empire, and a number of new talents exerted considerable influence on the development of these concepts. Perhaps the most illustrious and most famous of the ancient Islamic mathematicians was Muhammad ibn Musa al-Khwarizmi (780–850 CE). In 830 this scholar wrote an algebra text that became the definitive work in the subject.

Officially entitled *Al-Kitāb al-mukhtasar fi hīsāb al-jabr wa'l-muqabala* (which translates to *The Compendious Book on Calculation by Completion and Balancing*), the book is today known more concisely as *Kitab fi al-jabr wa'l-muqabala*; it introduced the now commonly used term "algebra" (from "al-jabr"). The word "jabr" referred to the balance maintained in an equation when the same quantity is added to both sides (the phrase "al-jabr" also came to mean "bonesetter"); the word "muqabala" refers to canceling like amounts from both sides of an equation.

Al-Khwarizmi's book *Art of Hindu Reckoning* introduced the notational system that we now call Arabic numerals: 1, 2, 3, 4, ... Our modern word "algorithm" was derived from a version of al-Khwarizmi's name, and some of his ideas contributed to the concept.

It is worth noting that good mathematical notation can make the difference between an idea that is clear and one that is obscure. The Arabs, like those who came before them, were hindered by lack of notation. When they performed their algebraic operations and solved their problems, they referred to everything with *words*. The scholars of this period are fond of saying that the Arabic notation was "rhetorical," with no symbolism of any kind. As an instance, we commonly denote the unknown in an algebraic equation by x; the Arabs would call that unknown "thing."

Moreover, the Arabs would typically exhibit their solutions to algebraic problems using geometric figures. They did not have an efficient method for simply writing the solution as we would today. It is clear that al-Khwarizmi had very clearly formulated ideas about algorithms for solving polynomial equations. But he did not have the notation to write the solutions down as we now do. He did not have the intellectual equipment (i.e., the formalism and the language) to formulate and prove theorems.

3.3 Investigations of Zero

The concept of zero has a long and colorful history. As early as 600 BCE, the Babylonians had a symbol in their arithmetic for zero. But they made it very clear that there were doubts about what "zero" actually meant. For how could a definite and explicit symbol stand for nothing?

The ancient Greeks, somewhat later, were obsessed with questions of the existence of matter and the essence of being. The concept of zero actually offended the Greeks.

In the Middle Ages, zero took on religious overtones. The concept of nothing seemed to have connections to the soul and to spirituality. In many contexts it was forbidden to discuss zero. People feared committing heresy. It was not until the sixteenth century that zero began to play a useful role in commerce. Obviously it will happen that a merchant will sell all units of product A. It is thus helpful to be able to write that "zero units of A remain in stock." Gradually, the idea of zero was incorporated into the standard mathematical argot. The objections faded away.

It seems fairly certain—according to the authoritative source [IFR]—that the zero concept has its origins with the Hindus in the early middle ages. The following outline, abbreviated from that source (and communicated to me by David H. Bailey), gives a feeling for the genesis of a number of related ideas:

Lokavibhaga, the Jaina cosmological text, appears using zero as a place-value.   458 CE

Aryabhata invents a method of recording numbers using zero and place-value.   510 CE

Astronomer and astrologer Varahamihra makes use of the place-value system
    with Sanskrit numerals.   575 CE

Arithmetician Jinabhadra Gani uses numerical expressions with a place-value
    system. $\sim$ 590 CE

Copper donation charter of Dadda II of Sankheda in Gujarat exhibits nine
    numerals in a place-value system.   594 CE

Sanskrit inscription of Cambodia exhibits word-symbols using the place-value system.	598 CE
Islamic mathematician al-Khwarizmi uses arabic numerals, including zero.	$\sim$ 820 CE
Gerbert of Aurillac learns the "Arabic system" of place-value in Islamic Spain.	$\sim$ 980 CE
Gerbert introduces Arabic numerals in Europe in Reims.	$\sim$ 990 CE

So it was Gerbert of Aurillac who introduced the arabic system of numerals and place-value in Europe. His efforts met with fierce resistance. Both for cultural and religious reasons, the Christians were wedded to the great Roman tradition, including the notation of Roman numerals. Even after Gerbert's death, he was accused of having been an alchemist and a sorcerer, and to have sold his soul to Lucifer.

The Crusades of Richard Lionheart did a great deal to bring the cultural riches of the Holy Land back to Europe, thus accomplishing much of what Gerbert's insight and energy failed to do. Leonardo of Pisa (1170–1250), better known as Fibonacci, visited Islamic North Africa and the Middle East. There he learned Arabic arithmetic, the operational techniques, the rules of algebra, and the basic ideas of Arabic geometry. Fibonacci's book *Liber abaci* (*The Book of the Abacus*) lays out many of these key ideas for a European audience. In those times many of the established experts carried out their craft on an abacus[36]—these were the adherents of the traditional Roman numerals—and Fibonacci's title was chosen in part for political reasons, to make peace with the abacists.

The Arabs referred to zero as *sifr*, meaning "empty." This is a translation of the Sanskrit "Shunya." When the concept arrived in Europe, people adopted the homophone *zephyrus*, meaning "west wind." In his text *Liber abaci*, Fibonacci uses the term *zephirum*. The Italian cognate is *zefiro*. This has led to our current usage *zero*.

Finally, after the French revolution the abacus was banned in schools and the Arabic numeral system with place-value was completely embraced.

As one might imagine, there were similar but perhaps more vigorous objections to the negative numbers. Over time, it was realized that negative numbers could play a useful role in commerce (as when a merchant is in debt), so negative numbers were gradually incorporated into the mathematical canon.

Today we use zero and the negative numbers with comfort and ease. We can actually *construct* number systems containing 0 and -5, so there is no longer any mystery as to where these numbers come from. Also the religious questions have faded into the background. So there remain few objections to these extended concepts of number.

It may be said that mathematical abstraction, and *proof*, have actually helped in the acceptance of the concepts of zero and the negative numbers. In the middle ages, when we could not say how these mysterious numbers arose, or precisely what they meant, these numbers had an extra air of mystery attached to them. Much thought of the time was influenced by religion, and people feared that things they did not understand were the

---

[36] See Section 7.1 for the history of the abacus and of computing machines.

works of the devil. Also the notion of a symbol that stood for nothing raised specters of evil signs and works of Satan. At various times, and by various people, it was actually *forbidden* to give explicit mention to zero or negative numbers. They were sometimes referred to explicitly in print as "forbidden" or "evil." The book [KAP1] gives a colorful history of the concept of zero. The authors demonstrate artfully that zero is an important and sometimes controversial concept, and one that has had considerable influence over human thought.

As we have indicated, mathematical theory put zero on a solid footing. It is possible to use the theory of equivalence classes (see [KRA5]) to *construct* the integer number system—in effect to *prove* that it exists, and that it has all the expected properties. So it becomes clear that this number system is a construct of human intellect, not of the devil or of some evil force.[37] Since the construction involves only pure thought, and follows the time-tested rigorous lines of logic that are typical of the mathematical method, there is a cleansing process that makes the result palatable and non-threatening. It is a triumph for mathematics and its culture.

3.4   The Idea of Infinity

The historical struggles with infinity were perhaps more strident and fear-laden than those for zero. Again, the reasons were religious. People felt that thinking about infinity was tantamount to thinking about the Creator. If one actually endeavored to *manipulate* infinity as a mathematical object—as we routinely manipulate numbers and variables and other constructs in mathematics—then one was showing the utmost disrespect, and exhibiting a cavalier attitude toward the Deity. Even highly educated industrials feared being guilty of heresy or sacrilege.

Many prominent nineteenth century mathematicians strictly forbade any discussion of infinity. Discussions of infinity were fraught with paradoxes and apparent contradictions, suggesting deep flaws in the foundations of mathematics. Such ideas only exacerbated people's fears and uncertainties. Today we have a much more sophisticated view of our discipline. We understand much more about the foundations of mathematics, and we have the tools for addressing (apparent) paradoxes and inconsistencies. In the nineteenth century, mathematicians were not so equipped. They felt quite helpless when mathematics revealed confusions or paradoxes or lacunae; so they shied away from such disturbing phenomena. The concept of infinity was the source of much misunderstanding and confusion. It was a subject best avoided. Fear of religious fallout gave a convenient rationale for pursuing such a course.

In the late nineteenth century mathematician Georg Cantor (1845–1918) finally determined the way to tame the infinity concept. He actually showed that there are many different "levels" or "magnitudes" of infinity. And he was able to prove some strikingly dramatic

---

[37] It is worth pondering whether the existence of 0 is a Platonic or a Kantian issue—see Section 1.7.

results—about classical topics like the transcendental numbers[38]—using his ideas about infinity.

Cantor suffered vehement and mean-spirited attacks—even from his senior associate Leopold Kronecker (1823–1891)—over his ideas about infinity. Cantor in fact had a very complicated relationship with Kronecker, and it is not entirely clear how great was Kronecker's role in Cantor's problems. Today there is evidence that Cantor suffered from bipolar depression. Also, Kronecker died 27 years before Cantor.

Cantor spent considerable time in asylums in an effort to cope with this calumny, and to deal with his subsequent depression. Near the end of his life, Cantor became disillusioned with mathematics; he spent a good portion of his final days in an effort to prove that Francis Bacon wrote the works of Shakespeare.

It is safe to say that Cantor's notions about infinity—his concept of *cardinality*, and his means of stratifying infinite sets of different orders or magnitudes—have been among the most profound and original ideas ever to be created in mathematics. These ideas have been universally embraced today and are part of the bedrock of modern mathematics. But in Cantor's day the ideas were most controversial. Originality comes at a price, and Cantor was flying in the face of severe prejudice and fears founded in religious dogma. People capitalized on the perception that Cantor was a Jew, and many of their attacks were shamelessly antisemitic—even though Cantor was in fact *not* a Jew. It is easy to see how Cantor might have become depressed, discouraged, and unwilling to go on.

Toward the end of his life, Kronecker came around and began to support Cantor and his ideas. Other influential mathematicians also began to see the light, and to support the program for the study of infinity. But by then it was virtually too late (at least for poor Georg Cantor). Cantor was a broken man. He received full recognition and credit for his ideas after his death; but in life he was a tormented soul.

---

[38] A real number is *algebraic* if it is the root of a polynomial equation with integer coefficients. A real number is *transcendental* if it is not algebraic. For example, the number $\sqrt{2}$ is algebraic because it is a root of the polynomial equation $x^2 - 2 = 0$. The number π is *not* algebraic—it is transcendental—because it is not the root of any such polynomial equation. This last assertion is quite difficult to prove.

4

The Dawn of the Modern Age

Every age has its myths and calls them higher truths.

—Anonymous

I have resolved to quit only abstract geometry, that is to say, the consideration of questions that serve only to exercise the mind, and this, in order to study another kind of geometry, which has for its object the explanation of the phenomena of nature.

—René Descartes

There is nothing new to be discovered in physics now. All that remains is more and more precise measurement.

—Lord Kelvin (1900)

It is now apparent that the concept of a universally accepted, infallible body of reasoning— the majestic mathematics of 1800 and the pride of man—is a grand illusion. Uncertainty and doubt concerning the future of mathematics have replaced the certainties and complacency of the past. The disagreements about the foundations of the "most certain" science are both surprising and, to put it mildly, disconcerting. The present state of mathematics is a mockery of the hitherto deep-rooted and widely reputed truth and logical perfection of mathematics.

—Morris Kline

A good proof is one that makes us wiser.

—Yuri Manin

4.1   Euler and the Profundity of Intuition

Leonhard Euler (1707–1783) was one of the greatest mathematicians who ever lived. He was also one of the most prolific. His collected works comprise more than 70 volumes, and they are still being edited today. Euler worked in all parts of mathematics, as well as mechanics, physics, and many other parts of science. He did mathematics almost effortlessly, often while *dangling* a grandchild on his knee. Late in life he became partially blind, but he declared that this would help him to concentrate more effectively, and his scientific output actually *increased*.

Euler exhibited a remarkable combination of mathematical precision and mathematical intuition. He had some of the deepest insights in all of mathematics, and he also committed some of the most famous blunders. One of these involved calculating the infinite sum

$$1 - 1 + 1 - 1 + 1 - + \cdots .$$

In fact, the *correct* way to analyze this sum is to add the first several terms and see whether a pattern emerges. Thus

$$1 - 1 = 0$$
$$1 - 1 + 1 = 1$$
$$1 - 1 + 1 - 1 = 0$$
$$1 - 1 + 1 - 1 + 1 = 1$$

and so forth. These "initial sums" are called *partial sums*. If the partial sums fall into a pattern and tend to some limit, then we say that the original infinite sum (or *series*) converges. Otherwise, we say that it diverges. What we see is that the partial sums for this particular series alternate between 0 and 1. They do not tend to any single, unique limit. So the series diverges.

But Euler did not know the correct way to analyze series (nor did anyone else in his day). His analysis went like this:

If we group the terms of the series as

$$(1 - 1) + (1 - 1) + (1 - 1) + \cdots$$

then the sum is clearly $0 + 0 + 0 + \cdots = 0$. But if we group the terms of the series as

$$1 + (-1 + 1) + (-1 + 1) + (-1 + 1) + \cdots$$

then the sum is clearly $1 + 0 + 0 + 0 + \cdots = 1$. Euler's conclusion was that this is a paradox.[39]

Euler is remembered today for profound contributions to number theory, geometry, the calculus of variations, complex analysis, and dozens of other fields in mathematics and the physical sciences. His blunders are all but forgotten. It was his willingness to take risks and to make mistakes that made him such an effective mathematician.

And this last is a point worth pondering. A mathematician is an explorer, and usually an explorer who has no idea what he/she is looking for. It is part of the ordinary course of life to make mistakes, to spend days on calculations that come to no conclusion, to pursue paths that end up being meaningless. But if one continues to analyze and to think critically and to stare mercilessly at all these efforts, then one may draw useful conclusions. In the end, one may formulate a theorem. And then, with an additional huge expenditure of hard work, one may prove that theorem. It is a great adventure, with many pitfalls and missteps. But that is the life of a mathematician.

---

[39] It may be noted that Euler also used specious argumentation to demonstrate that $1 + 2 + 4 + 8 + \cdots = -1$.

4.2 Dirichlet and the Heuristic Basis for Rigorous Proof

Peter Gustav Lejeune Dirichlet (1805–1859) was one of the great number theorists of the nineteenth century. His father's first name was Lejeune, coming from "Le jeune de Richelet," which means "young man from Richelette." The Dirichlet family came from the town of Liège in Belgium. The father was postmaster of Düren, the town where Peter was born. At a young age Dirichlet developed a passion for mathematics; he spent his pocket money on mathematics books. He entered the gymnasium[40] in Bonn at the age of 12, where he was a model pupil, exhibiting an interest in history as well as mathematics.

After two years at the *gymnasium*, Dirichlet's parents decided that they would rather have him at the Jesuit college in Cologne. There he fell under the tutelage of the distinguished scientist Georg Ohm (1787–1854). By age sixteen, Dirichlet had completed his school work and was ready for the university. German universities were not very good, nor did they have very high standards, at the time, so Peter Dirichlet decided to study in Paris. It is worth noting that several years later the German universities would set the worldwide standard for excellence; Dirichlet himself would play a significant role in establishing this preeminence.

Dirichlet always carried with him a copy of Gauss's *Disquisitiones arithmeticae* (Gauss's masterpiece on number theory), a work that he revered and kept at his side, much as other people might keep the Bible. Thus he came equipped for his studies in Paris, but he became ill soon after his arrival. However he would not let this deter him from attending lectures at the Collège de France and the Faculté des Sciences. He enjoyed the teaching of some of the leading scientists of the time, including Biot, Fourier, Francoeur, Hachette, Laplace, Lacroix, Legendre, and Poisson.

Beginning in the summer of 1823, Dirichlet lived in the house of the retired General Maximilien Sébastien Foy. He taught German to General Foy's wife and children. Dirichlet was treated very well by the Foy family, and he had time to study his mathematics.[41] During this period .... he published his first paper, which brought him instant fame. It dealt with Fermat's last theorem. The problem is to show that the Diophantine[42] equation

$$x^n + y^n = z^n$$

has no integer solutions x, y, z when n is a positive integer greater than 2. The cases $n = 3, 4$ had already been solved by Euler and by Fermat himself. Dirichlet decided to attack the case $n = 5$. This case divides into two subcases, and Dirichlet was able to dispatch subcase 1. Legendre was a referee of the paper, and he was able, after reading Dirichlet's work, to treat subcase 2. Thus a paper was published in 1825 that completely settled the case $n = 5$ of Fermat's last theorem. Dirichlet himself was subsequently able to develop his own proof of subcase 2 using an extension of his techniques for subcase 1. Later on, Dirichlet was also able to treat the case $n = 14$.

---

[40]Perhaps a word of explanation is in order here. In the United States, the *gymnasium* is a place where one plays basketball and works out on the parallel bars. In Germany, the *gymnasium* is the classical higher or secondary school for gifted students. Students are admitted at 10 or 11 years of age and (in West Germany) spend 9 years of study. The last year of the *gymnasium* is comparable to our freshman year in college.

[41]Also the Foys introduced Dirichlet to Fourier, thus sparking Dirichlet's lifelong passion for Fourier series.

[42]A *Diophantine equation* is an algebraic equation for which we seek whole number solutions.

██

Diophantine Equations

A *Diophantine equation* is an equation, usually with integer coefficients, for which we seek integer solutions. Perhaps the most famous Diophantine equation of all time is the Pythagorean equation

$$x^2 + y^2 = z^2 .$$

Elsewhere in this book we discuss how this equation arises from the study of right triangles. There are infinitely many triples of solutions to this equation.

Another famous Diophantine equation is Fermat's equation

$$x^k + y^k = z^k$$

for k an integer greater than 2. Fermat claimed to have an elegant proof that there are no solutions to this equation, but no proof was known for more than three hundred years. Finally, near the end of the twentieth century, Andrew Wiles and his student Richard Taylor proved that this Diophantine equation indeed has no solutions. They used modern techniques that were far beyond the capability of Fermat.

██

General Foy died in November 1825 and Dirichlet decided to return to Germany. However, in spite of support from Alexander von Humboldt, he could not assume a position in a German university since he had not submitted his *habilitation*. Dirichlet's mathematical achievements were adequate for such a thesis, but he was not allowed to submit because (**a**) he did not hold a doctorate and (**b**) he did not speak Latin.

The University of Cologne interceded and awarded Dirichlet an honorary doctorate. He submitted his *habilitation* on polynomials with prime divisors and obtained a position at the University of Breslau. Dirichlet's appointment was still considered controversial, and there was much discussion among the faculty of the merits of the case.

Standards at the University of Breslau were still rather low, and Dirichlet was not satisfied with his position. He arranged to transfer, again with Humboldt's help, to the Military College in Berlin. He also had an agreement that he could teach at the University of Berlin, which was one of the premier institutions of the time. Eventually, in 1828, he obtained a regular professorship at the University of Berlin, where he taught until 1855. Since he retained his position at the Military College, he was saddled with an unusual amount of teaching and administrative duties.

Dirichlet also earned an appointment at the Berlin Academy in 1831. His improved financial circumstances then allowed him to marry Rebecca Mendelssohn, the sister of the noted composer Felix Mendelssohn. Dirichlet obtained an eighteen-month leave from the University of Berlin to spend time in Italy with Carl Jacobi (who was there for reasons of his health).

In 1845, Dirichlet returned to his duties at the University of Berlin and the Military College. He continued to find his duties at both schools to be a considerable burden, and complained to his student Kronecker. It was quite a relief when, on Gauss's death in 1855, Dirichlet was offered Gauss's distinguished chair at the University in Göttingen.

Dirichlet endeavored to use the new offer as leverage to obtain better conditions in Berlin. But that was not to be, and he moved to Göttingen directly. There he enjoyed a quieter life with some outstanding research students. Unfortunately, the new blissful conditions were not to be enjoyed for long. Dirichlet suffered a heart attack in 1858, and his wife died of a stroke shortly thereafter.

Dirichlet's contributions to mathematics were monumental. We have already described some of his work on Fermat's last theorem. He also made contributions to the study of Gauss's quadratic reciprocity law. It can be said that Dirichlet was the father of the subject of analytic number theory. In particular, he proved foundational results about prime numbers occurring in arithmetic progression.[43]

Dirichlet had a powerful intuition, which guided his thoughts decisively as he developed his mathematics. But he was widely recognized for the precision of his work. No less an eminence than Carl Jacobi (1804–1851) said,

> If Gauss says he has proved something, it seems very probable to me; if Cauchy [Augustin-Louis Cauchy (1789–1857)] says so, it is about as likely as not; if Dirichlet says so, it is certain. I would gladly not get involved in such delicacies.

Dirichlet did further work on what was later to become (in the hands of Emmy Noether) the theory of *ideals*. He created *Dirichlet series*, which are today a powerful tool for analytic number theorists. And he laid some of the foundations for the theory of *class numbers* (later to be developed by Emil Artin).

Dirichlet is remembered for giving one of the first rigorous definitions of the concept of function. He was also among the first to define—at least in the context of Fourier series— precisely what it means for a series to *converge* (Cauchy dealt with this issue somewhat earlier). He is remembered as one of the fathers of the theory of Fourier series.

Dirichlet had a number of historically important students, including Kronecker and Riemann. Riemann went on to make seminal contributions to complex variables, Fourier series, and geometry.

---

[43]Only recently, in [GRT], Benjamin Green and Terence Tao proved that there are arbitrarily long arithmetic progressions of prime numbers. An arithmetic progression is a sequence of numbers that are equally spaced apart. For example,

$$3 \quad 5 \quad 7$$

is a sequence of three primes, spaced 2 apart. This is an arithmetic progression of primes of *length 3*. As a second example,

$$11 \quad 17 \quad 23 \quad 29$$

is a sequence of four primes spaced 6 apart. This is an arithmetic progression of *length 4*. See whether you can find an arithmetic progression of primes of length 5. Suffice it to say that Green and Tao used very abstract methods to establish that there are arithmetic progressions of primes of any length.

4.2.1 The Pigeonhole Principle

Today combinatorics, number theory, and finite mathematics are thriving enterprises. Cryptography, coding theory, queuing theory, and theoretical computer science all make use of counting techniques. But the idea of "counting," as a science, is relatively new.

Dirichlet was one of the greatest workers in number theory. Many theorems and ideas in that subject are named after him. But he was in fact loathe to spend time finding rigorous proofs of his new discoveries. He generally proceeded with a keen intuition and a profound grasp of the main ideas. But he left it to others, and to future generations, to establish the results rigorously.

Dirichlet was one of the first masters of the theory of counting. One of his principal counting techniques, the one for which he is most vividly remembered, is that which was originally called the "Dirichletscher Schubfachschluss" or "Dirichletscher Schubfachprincip" (Dirichlet's drawer shutting principle). Today we call it the "pigeonhole principle." It is a remarkably simple idea that has profound consequences. The statement is this:

> If you put $n + 1$ letters into n mailboxes then some mailbox will contain at least two letters.

This is quite a simple idea (though it *can* be given a rigorous mathematical proof—see Chapter 12). It says, for instance, that if you put 101 letters into 100 mailboxes then some mailbox will contain at least two letters. The assertion makes good intuitive sense. This pigeonhole principle turns out to be a terrifically useful mathematical tool. As a simple instance, if you have 13 people in a room, then at least two of them will have their birthdays in the same month. Why? Think of the months as mailboxes and the birthdays as letters.

Dirichlet in fact applied his pigeonhole principle to prove deep and significant facts about number theory. Here is an important result that bears his name, and is frequently cited today:

> **Theorem.** *Let ξ be a real number. If $n > 0$ is an integer, then there are integers p, q such that $0 < q \le n$ and*
>
> $$\left| \frac{p}{q} - \xi \right| < \frac{1}{nq} .$$

A key idea in number theory is that irrational numbers and transcendental numbers may be characterized by the rate at which they can be approximated by rational numbers. This entire circle of ideas originates with Dirichlet's theorem, and the root of that theorem is the simple but profound pigeonhole principle.

4.3 The Golden Age of the Nineteenth Century

Nineteenth century Europe was a haven for brilliant mathematics. So many of the important ideas in mathematics today grew out of ideas that were developed at that time. We list just a few of these:

- Jean Baptiste Joseph Fourier (1768–1830) developed the seminal ideas for Fourier series and created the first formula for the expansion of an arbitrary function into a trigonometric series. He developed applications to the theory of heat.
- Évariste Galois (1812–1832) and Augustin-Louis Cauchy laid the foundations for abstract algebra by inventing group theory.
- Bernhard Riemann (1826–1866) established the subject of differential geometry, defined the version of the integral (from calculus) that we use today, and made profound contributions to complex variable theory and Fourier analysis.
- Augustin-Louis Cauchy laid the foundations of real analysis, complex variable theory and partial differential equations. He also did seminal work in geometric analysis.
- Carl Jacobi (1804–1851), Ernst Kummer (1810–1893), Niels Henrik Abel (1802–1829), and numerous other mathematicians from many countries developed number theory.
- Joseph-Louis Lagrange (1736–1813), Cauchy, and others laid the foundations of the calculus of variations, classical mechanics, the implicit function theorem, and many other important ideas in modern geometric analysis.
- Karl Weierstrass (1815–1897) laid the foundations for rigorous analysis with numerous examples and theorems. He made seminal contributions both to real and to complex analysis.

This list could be expanded considerably. The nineteenth century was a fecund time for European mathematics, and communication among mathematicians was at an all-time high. There were several prominent mathematics journals, and important work was widely disseminated. The many great universities in Italy, France, Germany, and England (England's were driven by physics) had vigorous mathematics programs and many students. This was an age when the foundations for modern mathematics were laid.

And the seeds of rigorous discourse were being sown at this time. The language and terminology and notation of mathematics were not quite yet universal, the definitions were not well established, and even the methods of proof were in development. But the basic methodology was in place and the mathematics of that time traveled reasonably well among countries and to the twentieth century and beyond. As we shall see below, Bourbaki and Hilbert set the tone for rigorous mathematics in the twentieth century. But the work of the many nineteenth century geniuses paved the way for those pioneers.

5

Hilbert and the Twentieth Century

There is no religious denomination in which the misuse of metaphysical expressions has been responsible for so much sin as it has in mathematics.

—Ludwig Wittgenstein

Mathematics is not necessarily characterized by rigorous proofs. Many examples of heuristic papers written by prominent mathematicians are given in [JAQ]; one can list many more papers of this kind. All these papers are dealing with mathematical objects that have a rigorous definition.

—Albert Schwarz

The most vexing of issues ... is the communication of insights. Unlike most experimentalized fields, Mathematics does not have a vocabulary tailored to the transmission of condensed data and insight. As in most physics experiments, the amount of raw data obtained from mathematical experiment will, in general, be too large for anyone to grasp. The collected data need to be compressed and compartmentalized.

—J. Borwein, P. Borwein, R. Girgensohn, and S. Parnes

It is by logic we prove, it is by intuition that we invent.

—Henri Poincaré

5.1 David Hilbert

Along with Henri Poincaré (1854–1912) of France, David Hilbert (1862–1943) of Germany was the spokesman for early twentieth century mathematics. Hilbert is said to have been one of the last mathematicians to be conversant with the entire subject—from differential equations to analysis to geometry to logic to algebra. He exerted considerable influence over all parts of mathematics, and he wrote seminal texts in many of them. Hilbert had an important and profound vision for the rigorization of mathematics (one that was later dashed by the work of Bertrand Russell, Kurt Gödel, and others), and he set the tone for the way that mathematics was to be practiced and recorded in our time.

Abstract Algebra

In an earlier part of the book we encountered the concept of "group." This is just one of many abstract algebraic structures that exemplifies certain relationships among objects. A *ring* is a collection of objects with two operations for combining them—typically addition and multiplication. There are rings of numbers, rings of matrices, rings of operators on a Hilbert space, and many other examples as well. A *field* is a ring that has a notion of division.

Abstract algebraists study these different algebraic structures and their mutual relations. Today abstract algebra has applications in many fields of mathematics, physics, and engineering.

Hilbert had many important students, ranging from Richard Courant (1888–1972) to Theodore von Kármán (1881–1963) (the father of modern aeronautical engineering) to Hugo Steinhaus (1887–1972) to Hermann Weyl (1885–1955). His influence was felt widely, not just in Germany but around the world. He was the primary force in establishing Göttingen as one of the world centers for mathematics, and it is still an important center today.

One of Hilbert's real coups was to study the subject of algebraic invariants. As a simple example, let

$$p(x) = ax^2 + bx + c$$

be a quadratic polynomial. One of the most basic invariants of such a polynomial is the *discriminant*

$$d = b^2 - 4ac.$$

If the variable x is subjected to a linear transformation then, up to scaling, the discriminant does not change. One would like to classify all algebraic invariants.

A good deal of late nineteenth century mathematics was dedicated to efforts to actually record explicitly all the invariants of different types. Many mathematicians dedicated their entire careers to the effort. Another way of putting this is that people tried to write down a *basis* for the invariants.[44] Hilbert established the result nonconstructively, essentially with a proof by contradiction. This was highly controversial at the time—even though proof by contradiction had been accepted by many mathematician at least since the time of Euclid. Hilbert's work put a great many mathematicians out of business, and established him rather quickly as a force to be reckoned with. His ideas essentially created the subject of abstract algebra. Hilbert is remembered today for a great many mathematical innovations, one of which was his *Nullstellensatz*—one of the key algebraic tools that he developed for the study of invariants.

---

[44] A "basis" is a minimal generating set for an algebraic system.

5.2 G.D. Birkhoff, Norbert Wiener, and the Development of American Mathematics

In the late nineteenth century and early twentieth century, American mathematics had something of a complex. Not a lot of genuine (abstract, rigorous) mathematical research was being done in the U.S.A., and the preeminent mathematicians of Europe—the leaders in the field—looked down their collective noses at the paltry American efforts. We all must grow where we are planted, and the American intellectual life was a product of its context. America was famous then, even as it is now, for being practical, empirical, rough-and-ready, and eager to embrace the next development—whatever it may be. America prides itself on being a no-holds-barred society, in which there is great social mobility and few if any obstacles to progress. Intellectual life in the nineteenth century United States reflected those values. University education was concrete and practical, and grounded in particular problems that came from engineering or other societal needs.

As a particular instance of the point made in the last paragraph, William Chauvenet (1820–1870) was one of the intellectual mathematical leaders of nineteenth century America. He was one of the founders of the U. S. Naval Academy in Annapolis, and then he moved to Washington University in St. Louis (home of this author). In those days most American universities did not have mathematics departments. In fact, generally speaking, mathematics was part of the astronomy department. This fit rather naturally, from a practical point of view, because astronomy involves a good deal of calculation and mathematical reckoning. Chauvenet, however, convinced the administration of Washington University to establish a freestanding mathematics department. And he was its first chairman. Chauvenet went on to become the chancellor of Washington University, and he played a decisive role in the institution's early development.

So what kind of mathematical research did a scholar like Chauvenet do? Perhaps the thing he is most remembered for is that he did all the calculations connected with the construction of the Eads Bridge (which still spans the Mississippi River in downtown St. Louis). The Eads Bridge is a great example of the classical arch style of bridge design, and it still stands and is in good use today—both as a footbridge and for automobile traffic.

The premier mathematicians of Europe—Riemann and Weierstrass and Dirichlet and Gauss in Germany as well as Cauchy and Liouville and Hadamard and Poincaré in France—were spending their time developing the foundations of real analysis, complex analysis, differential geometry, abstract algebra, and other fundamental parts of modern mathematics. You would not find them doing calculations for bridges.[45] There was a real disconnect between European mathematics and American mathematics.

One of the people who bridged the gap between the two mathematical cultures was J.J. Sylvester. It happened that Johns Hopkins University in Baltimore was seeking a new

---

[45]To be fair, Gauss—especially later in his career—took a particular interest in a variety of practical problems. A number of other European mathematicians, famous for their abstract intellectual work, also did significant applied work. But it is safe to say that their roots and their focus were in pure mathematics. This is the work that has stood the test of time, and for which they are remembered today.

mathematical leader, and the British mathematician Sylvester was brought to their attention. In fact it came about in this way:

When Harvard mathematician Benjamin Peirce (1809–1880) heard that the Johns Hopkins University was to be founded in Baltimore, he wrote to the new president Daniel Coit Gilman (1831–1908) in 1875 as follows:

> Hearing that you are in England, I take the liberty to write you concerning an appointment in your new university, which I think would be greatly for the benefit of our country and of American science if you could make it. It is that of one of the two greatest geometers of England, J.J. Sylvester. If you enquire about him, you will hear his genius universally recognized but his power of teaching will probably be said to be quite deficient. Now there is no man living who is more luminary in his language, to those who have the capacity to comprehend him than Sylvester, provided the hearer is in a lucid interval. But as the barn yard fowl cannot understand the flight of the eagle, so it is the eaglet only who will be nourished by his instruction . . . . Among your pupils, sooner or later, there must be one, who has a genius for geometry. He will be Sylvester's special pupil—the one pupil who will derive from his master, knowledge and enthusiasm—and that one pupil will give more reputation to your institution than the ten thousand, who will complain of the obscurity of Sylvester, and for whom you will provide another class of teachers . . . . I hope that you will find it in your heart to do for Sylvester— what his own country has failed to do—place him where he belongs—and the time will come, when all the world will applaud the wisdom of your selection.

Sylvester (1814–1897) was educated in England. When he was still young, he accepted a professorship at the University of Virginia. One day a young student whose classroom recitation Sylvester had criticized became quite piqued with the esteemed scholar. He prepared an ambush and fell upon Sylvester with a heavy walking stick.[46] Sylvester speared the student with a *sword cane*, which he just happened to have handy. The damage to the student was slight, but the professor found it advisable to leave his post and take the earliest possible ship to England. Sylvester took a position there at a military academy. He served long and well, but subsequently retired and accepted a position, when he was in his late fifties, at Johns Hopkins University. He founded the *American Journal of Mathematics* the following year, and was the leading light of American mathematics in his day.

H.F. Baker (1866–1956) recounts the following history of Sylvester and the Johns Hopkins University:

> In 1875 the Johns Hopkins University was founded at Baltimore. A letter to Sylvester from the celebrated Joseph Henry (1797–1878), of date 25 August 1875, seems to indicate that Sylvester had expressed at least a willingness to share in forming the tone of the young university; the authorities seem to have felt that a Professor of Mathematics and a Professor of Classics could inaugurate the work of a University without expensive buildings or elaborate apparatus. It was finally agreed that Sylvester should go, securing besides his travelling expenses, an annual stipend of 5000 dollars "paid in gold." And so, at the age of sixty-one,

---

[46]Today professors sometimes complain about their students' rude behavior. But they do not often mention being attacked by a student with a walking stick.

still full of fire and enthusiasm ... he again crossed the Atlantic, and did not relinquish the post for 8 years, until 1883. It was an experiment in educational method; Sylvester was free to teach whatever he wished in the way he thought best; so far as one can judge from the records, if the object of any University be to light a fire of intellectual interests, it was a triumphant success. His foibles no doubt caused amusement, his faults as a systematic lecturer must have been a sore grief to the students who hoped to carry away note books of balanced records for future use; but the moral effect of such earnestness ... must have been enormous.

Sylvester once remarked that Arthur Cayley had been much more fortunate than himself: "that they both lived as bachelors in London, but that Cayley had married and settled down to a quiet and peaceful life at Cambridge; whereas he [Sylvester] had never married, and had been fighting the world all his days." Those in the know attest that this is a fair summary of their lives.

Teaching is an important pursuit, and has its own rewards. So is research. But the two can work symbiotically together, and the whole created thereby is often greater than the sum of its parts. Sylvester, who was an eccentric teacher at best, describes the process in this way:

But for the persistence of a student of this university in urging upon me his desire to study with me the modern algebra I should never have been led into this investigation; and the new facts and principles which I have discovered in regard to it (important facts, I believe), would, so far as I am concerned, have remained still hidden in the womb of time. In vain I represented to this inquisitive student that he would do better to take up some other subject lying less off the beaten track of study, such as the higher parts of the calculus or elliptic functions, or the theory of substitutions, or I wot [know] not what besides. He stuck with perfect respectfulness, but with invincible pertinacity, to his point. He would have the new algebra (Heaven knows where he had heard about it, for it is almost unknown in this continent [America]), that or nothing. I was obliged to yield, and what was the consequence? In trying to throw light upon an obscure explanation in our text-book, my brain took fire, I plunged with re-quickened zeal into a subject which I had for years abandoned, and found food for thoughts which have engaged my attention for a considerable time past, and will probably occupy all my powers of contemplation advantageously for several months to come.

Sylvester once gave a commencement address at Johns Hopkins. He began by remarking that mathematicians were not any good at public speaking because the language of mathematics was antithetical to general communication. That is, mathematics is concise: one can express pages of thought in just a few symbols. Thus, since he was accustomed to mathematical expression, his comments would be painfully brief. He finished three hours later.

Sylvester once sent a paper to the London Mathematical Society for publication. True to form, he included a cover letter asserting that this was the most important result in the subject for 20 years. The secretary replied that he agreed entirely with Sylvester's assessment, but that Sylvester had actually published the result in the *Journal of the London Mathematical Society* five years earlier.

Suffice it to say that Sylvester was the most powerful player in sight in American mathematics for a good many years. He set a sterling example for scholarship, an influential teacher, and he accomplished much. But there was more to come.

The next big event, from the point of view of American mathematics, is that G.D. Birkhoff (1884–1944) came along. Educated at Harvard, he stayed on and joined the faculty there. And he was the first native-born American to prove a theorem that caught the attention and respect of the European mandarins in the field. This was his proof of Poincaré's last theorem (Poincaré's proof had been considered unsatisfactory, and the theorem itself very important). Also his proof of the general ergodic theorem attracted considerable interest; it was a problem that had received broad and intense interest for many years. Birkhoff cracked it, and he thereby put himself, Harvard, and American mathematics on the map. Now America was a real player in the great mathematical firmament.

It took more than just one man to earn the undying respect and admiration of the rather stodgy European mathematicians. So we can be grateful that Norbert Wiener (1894–1964) came on the scene. Wiener was a child prodigy, tutored by his martinet father (who was himself an academic). Norbert was born on the campus of the University of Missouri in Columbia, where his father was a professor of languages. The elder Wiener was on the losing end of some pivotal political battle, and as a result was forced to leave his position in Missouri. The family ended up moving to the Boston area. After floundering around for a while, Wiener père ended up on the faculty at Harvard. He remained there for the rest of his career.

Young Norbert, who was given a running start in his education by his father's diligent attentions, began attending Tufts University at the age of eleven. He was at the time the youngest student in America ever to attend college. There was considerable press coverage of this event, and Norbert was an instant celebrity. Norbert learned when he was 17-years old that he was Jewish (prior to that his father had represented that the family was gentile). Wiener took this news very badly, and the fact of being Jewish plagued him for the rest of his life. He felt that anti-Semitic forces (G.D. Birkhoff notably among them) were in power in American mathematics. So at the start of his mathematical career, Wiener lived in England, where he felt he could get a fair shake. He was finally able, through some careful politicking, to land a job at MIT. Wiener was to spend the rest of his professional life in Cambridge, Massachusetts.

Wiener was short, rotund, and extremely myopic. He cut quite a figure as he strode around the MIT campus. But he was very famous for his intellectual prowess. His classes had to be held in great halls, because people attended from all over the Boston area. It was rumored that Wiener's salary was higher than that of the President of MIT.

Wiener put MIT and American mathematics (and himself, of course) on the map yet again. His work in Fourier analysis and stochastic integrals and cybernetics (a term, and a subject, that he invented) was groundbreaking, and his theorems are still studied and cited today.

From the point of view of mathematical proof, Norbert Wiener (like Sylvester and Birkhoff) was a classicist. He formulated theorems in the traditional way, and proved them rigorously with pen and paper. But Wiener also maintained a considerable interest in the *applications* of mathematics, and in the way that scientists interact with society. He was deeply troubled by the use of the atomic bomb to end World War II in Japan. He campaigned against scientists lending their intellect to support the military.

Norbert Wiener was one of the fathers of the modern theory of *stochastic processes*. This is a branch of probability theory that analytically describes random processes, such as Brownian motion. In the late 1940s and early 1950s, all of Norbert Wiener's many interests converged in such a way that he invented an entirely new avenue of human inquiry. He called it "cybernetics."[47]

Cybernetics is the study of how man interacts with machines (particularly, but not exclusively, computers). It considers questions such as whether a machine can "think." It turned out that these were not merely questions of philosophical speculation. Wiener could analyze them using his ideas from stochastic processes. He wrote copious technical papers on cybernetics, and traveled the world spreading his gospel. And these ideas enjoyed some real currency in the 1950s. Wiener had quite a following, and at MIT he had a cybernetics lab with some top scholars as his coworkers.

It is safe to say that G.D. Birkhoff and Norbert Wiener were two of the key players who helped thrust American mathematics into a prominent position in the twentieth century. They both had some truly important and influential students: Birkhoff's included Marston Morse (1892–1977), Hassler Whitney (1907–1989), and Marshall Stone (1903–1989); Wiener's included Amar Bose (of Bose Stereo fame), Norman Levinson, and Abe Gelbart.

The early twentieth century was a golden time for American mathematics, and for American scholarship in general. The opening of the Institute for Advanced Study in Princeton in 1930, with its galaxy of world-class mathematicians (and of course with superstar physicist and Nobel Laureate Albert Einstein) helped to put American mathematics into the fore. The University of Chicago, founded with money from John D. Rockefeller in 1890 (the first classes were held in 1892), also became a bastion of American mathematical strength. It was soon followed by Princeton and Harvard, and later by MIT (thanks to the role model and considerable effort of Norbert Wiener).

Today America is unquestionably one of the world leaders in mathematics. There are several reasons for this preeminence. First, America has many first-class universities. Secondly, there is considerable government subsidy (through the National Science Foundation, the Department of Defense, the Department of Energy, and many other agencies) for mathematical research. But it might be noted that the American way of doing business has played a notable role in the development of mathematics. In America, a hard-working and successful mathematician can really move ahead. Beginning with a humble Ph.D. and a modest job, if you go on to prove important theorems, you will get better job offers. It is definitely possible to move up through the ranks—to more and more prestigious positions and more elite universities—and to get a better salary and a fancy job with more perks. The top mathematicians in the United States have extremely good salaries, discretionary funds, a stable of assistants (postdocs and assistant professors and others who work with them), and many other benefits.

This is not how things are in most other countries. In many of the leading European countries, all education is centralized. In Italy all the decisions come out of Rome, and in

---

[47]The word *cybernetics* derives from the Greek *kybernētēs*, which means "steersman" or "helmsman."

France they all come from Paris. This has certain benefits—in terms of nationalized standards and uniformity of quality, but it also makes the system stodgy and inflexible. When Lars Hörmander, a Swede, won the Fields Medal in 1962 there was no job for him in Sweden and they were unable to create one for him. How can this be? It is a fact that, in Sweden, the total number of full professorships in mathematics is a fixed constant. Today it is twenty. But in 1960 it was nineteen. All 19 positions were filled, and nobody was about to step down or abandon his professorship so that Hörmander might have a job. [Recall that Isaac Newton's teacher Isaac Barrow did indeed give up his Lucasian chair so that the brilliant young Newton could assume it.][48] But nothing like this was going to happen in *socialistic* Sweden in 1962. So Hörmander, never a wilting flower, quit Sweden and moved to the Institute for Advanced Study in Princeton, New Jersey. This is arguably one of the most prestigious mathematics jobs in the world, with a spectacular salary and many lovely perks. So Hörmander flourished in his new venue. After several years, the Swedes became conscious of their loss. One day an insightful politician stood up in the national Parliament and said, "I think it's a tragedy that the most brilliant Swedish mathematician in history cannot find a suitable job in our country. He has left and moved to the United States." So in fact the Swedish parliament created one more mathematics professorship—raised the sacred total from 19 to 20—so that Hörmander could have a professorship in Sweden. And, loyal Swede that he was, Hörmander dutifully moved back to Sweden. He has been at the University of Lund for nearly 40 years.

But it should be noted that Hörmander's job in Lund (he is now retired) was nothing special. He took a more than 50% cut in pay to return to Sweden, and his salary was about the same as that of any other senior professor. And this is so because such things as professors' salaries are centrally regulated, and a system of central regulation does not take into account exceptional individuals like Lars Hörmander.

The point here is that the American system is more competitive—the academic world is really a marketplace—and that perhaps motivates people to strive harder and to seek greater heights.[49] By contrast, a German academic once told me that he knew what his salary would be in 5 years or 10 years or 20 years because the government had published a book laying it all out. If he got an outside offer, or moved to another university, then nothing would change. Everything was centrally regulated.

One of the chief tenets of American education is "local control of schools." This has its upsides and its downsides; education in rural Mississippi is quite different from education in Boston. But the American system contributes to a more competitive atmosphere. An American university (especially a private one) would have no trouble at all creating a special professorship for a scholar like Lars Hörmander. And it would not be a cookie-cutter position just like any other professorship at the university. The chancellor and the provost could tailor the position to fit the achievements and the prestige of the individual in question. They could cook up any salary they desired, and assign any perks to the position that they

[48] Barrow had other motivations for this largesse: he wanted to assume a position at court. Nonetheless, his resignation gave Newton the opportunity he needed.

[49] It must be noted, however, that until recently the Russian and the Romainian systems were highly impressive, and produced a myriad of brilliant mathematicians and great theorems. Under the communist regime, the Russian system was *not* a marketplace. But Russian culture is special in many ways, and there were other forces at play.

thought appropriate. The distinguished professor could have a private secretary, his own limousine, or whatever they thought would keep him happy and loyal to the institution.[50]

It is also the case that the American academic world is much less of a caste system than the European or Asian systems. Things are much less structured, and there is much more mobility. It is very common for American assistant professors, associate professors, and full professors to be friends. American professors will joke around with the secretaries and have lunch with the graduate students. One sees much less of this type of behavior at foreign universities. Perhaps the free and open nature of our system contributes to its success.

But it must be said that one of the keys to the success of the American academic system is money, pure and simple. Higher education is expensive, and American universities have access to many more resources than do European universities. There are many government and private funding agencies, and also there is a great tradition at American universities of alumni giving (not so in England, for example). Harvard University has an endowment that exceeds 26 billion dollars, and much of that money comes from alumni gifts. It is quite common for a wealthy donor to approach an American university with the offer of $10 million to endow five academic chairs.[51] This is a special academic position, with a fine salary and many benefits, that is designed to attract and retain the best scholars. Nothing like this ever happens in Europe.

Many brilliant scholars leave Europe to accept positions at American universities because the salaries are so much higher (and, concomitantly, the working conditions so much more conducive to productivity). At the same time, they often return to their homeland when they retire.

5.3 L.E.J. Brouwer and Proof by Contradiction

L.E.J. Brouwer (1881–1966) was a bright young Dutch mathematician whose chief interest was in topology. Now topology was quite a new subject in those days (the early twentieth century). Affectionately dubbed "rubber sheet geometry," the subject concerns itself with geometric properties of surfaces and spaces that are preserved under continuous deformation (i.e., twisting and bending and stretching). Brouwer established his reputation by making important contributions to the study of the Jordan curve theorem (the amazingly subtle result that says that a closed, non-self-intersecting curve in the plane divides the plane into two regions—the interior of the curve and the exterior of the curve). In his continuing studies of this burgeoning new subject, Brouwer came up with a daring new result, and he found a way to prove it.

---

[50]These days, at UCLA, the truly big star on the math faculty is Terence Tao. He has won the Fields Medal, the MacArthur Fellowship, the Waterman Award, the Bôcher Prize, and many other *awards*. The university wants to treat him right, so it has created a special professorship for Tao with a special salary. And they have also bought him a house in Bel Air, one of the *tonier* parts of town.

[51]For example, this type of situation occurred at Penn State in 1986. Wealthy donor Robert Eberly gave money to each of the science departments—biology, chemistry, astronomy, and physics—to establish an endowed chair. He declined to give money for mathematics because he had never had a math class at Penn State that he liked. But the university was able to use its own funds to repair that situation.

Known as the "Brouwer fixed-point theorem," the result can be described as follows. Consider the closed unit disk $\overline{D}$ in the plane, as depicted in Figure 5.1. This is a round, circular disk—including the boundary circle as shown in the picture. Now imagine a function $\varphi : \overline{D} \to \overline{D}$ that maps this disk continuously to itself, as shown in Figure 5.2. Brouwer's result is that the mapping φ must have a fixed point. That is to say, there is a point $P \in \overline{D}$ such that $\varphi(P) = P$. See Figure 5.3.

Figure 5.1. The closed disk.

This is a technical mathematical result, and its rigorous proof uses profound ideas such as homotopy. But, serendipitously, it lends itself rather naturally to some nice heuristic explanations. Here is one popular interpretation. Imagine that you are eating a bowl of soup—Figure 5.4. You sprinkle grated cheese uniformly over the surface of the soup (see Figure 5.5). And then you stir up the soup. We assume that you stir the soup in a civilized manner so that all the cheese remains on the surface of the soup (refer to Figure 5.6). Then some grain of cheese remains in its original position (Figure 5.7).

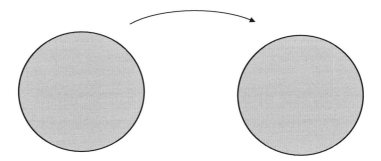

Figure 5.2. A continuous map of the disk.

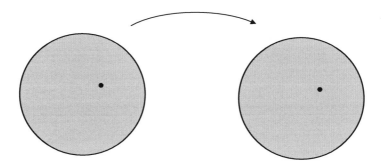

Figure 5.3. A fixed point.

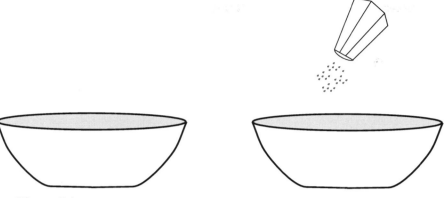

Figure 5.4. A bowl of soup.   **Figure 5.5.**   Grated cheese on the soup.

The soup analogy gives a visceral way to think about the Brouwer fixed-point theorem. Both the statement and the proof of this theorem—in the year 1909—were quite dramatic. In fact, it is now known that the Brouwer fixed-point theorem is true in every dimension (Brouwer himself proved it only in dimension 2). We shall provide some discursive discussion of the result below.

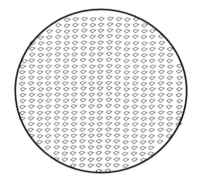

Figure 5.6.   The closed disk.

The Brouwer fixed-point theorem is one of the most fascinating and important theorems of twentieth century mathematics. Proving this theorem established Brouwer as one of the preeminent topologists of his day. But he refused to lecture on the subject, and in fact he ultimately rejected this (his own!) work. The reason for this strange behavior is that L.E.J. Brouwer had become a convert to *constructivism*, or *intuitionism*. He rejected the Aristotelian dialectic (that a statement is either true or false and there is no alternative), and therefore rejected the concept of "proof by contradiction." Brouwer had come to believe that the only valid proofs—at least when one is proving *existence* of some mathematical object (such as a fixed point!) and when infinite sets are involved—are those in which we *construct* the asserted objects being discussed.[52] Brouwer's school of thought became known as "intuitionism," and it has made a definite mark on twentieth century mathematics. No less an *éminence grise* than Hermann Weyl subscribed to parts of the intuitionist philosophy, and Errett Bishop (see below) defended it vigorously.

---

[52] In fact, for the constructivists, the phrase "there exists" must take on a rigorous new meaning that exceeds the usual rules of formal logic.

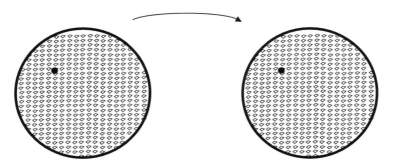

Figure 5.7. A fixed grain of cheese.

As we shall see below, the Brouwer fixed-point theorem asserts the existence of a "fixed point" for a continuous mapping. One demonstrates that the fixed point exists by assuming that it does not exist, thereby deriving a contradiction. This is Brouwer's original method of proof, but the methodology flies in the face of the intuitionism that he later adopted.

Let us begin by discussing the general idea of the Brouwer fixed-point theorem. We proceed by considering a "toy" version of the question in one dimension. Consider a continuous function f from the interval $[0, 1]$ to $[0, 1]$. Figure 5.8 exhibits the graph of such a function.

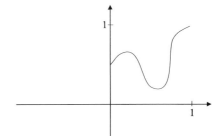

Figure 5.8. A continuous function on the unit interval.

Note here that the word "continuous" refers to a function that has no breaks in its graph. Some like to say that the graph of a continuous function "can be drawn without lifting the pencil from the paper." Although there are more mathematically rigorous definitions of continuity, this one will suffice for our purposes. The question is whether there is a point $p \in [0, 1]$ such that $f(p) = p$. Such a point p is called a *fixed point* for the function f. Figure 5.9 shows how complicated a continuous function from $[0, 1]$ to $[0, 1]$ can be. In each instance it is not completely obvious whether there is a fixed point. But in fact Figure 5.10 exhibits the fixed point in each case.

It is one thing to draw a few pictures and quite another to establish once and for all that no matter what the choice of the continuous function $f : [0, 1] \to [0, 1]$, there is a fixed point p. What is required now is a *mathematical proof*. Here is a formal enunciation and proof of our result:

Theorem 5.3.1. *Let* $f : [0, 1] \to [0, 1]$ *be a continuous function. Then there is a point* $p \in [0, 1]$ *such that* $f(p) = p$.

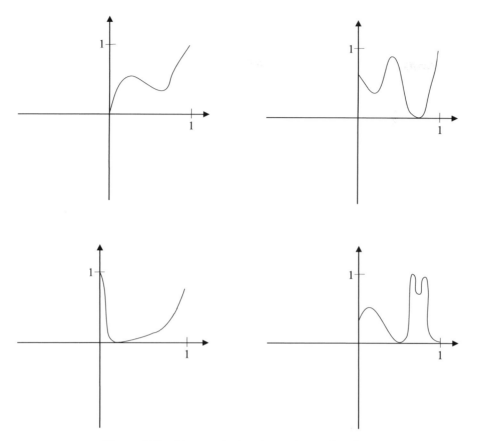

Figure 5.9. The complexity of a continuous function.

Proof. We may as well suppose that $f(0) \neq 0$ (otherwise 0 is our fixed point and we are done). Thus $f(0) > 0$. We also may as well suppose that $f(1) \neq 1$ (otherwise 1 is our fixed point and we are done). Thus $f(1) < 1$.

Consider the auxiliary function $g(x) = f(x) - x$. By the observations in the last paragraph, $g(0) > 0$ and $g(1) < 0$. Look at Figure 5.11. We see that a continuous function with these properties must have a point p in between 0 and 1 such that $g(p) = 0$. But this just says that $f(p) = p$.[53]   □

Now we turn to the higher-dimensional, particularly the 2-dimensional, version of the Brouwer fixed-point theorem. This is the formulation that caused such interest and

---

[53] We are using here the important intermediate value property of a continuous function. This theorem says that if a continuous function takes the values α and β, with $\alpha < \beta$, then it must take all the values in between. This key fact is bound up with the axioms of the real number system, and the completeness of that system. A thorough treatment appears, for example, in [KRA5] and [KRA7].

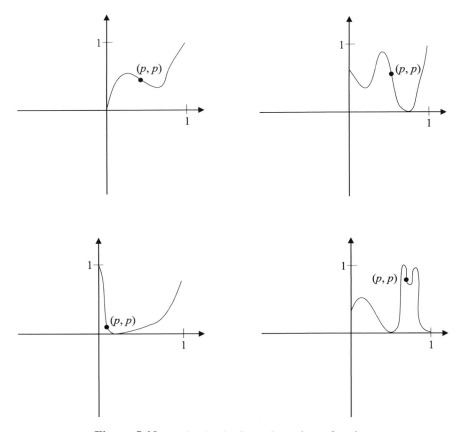

Figure 5.10.   A fixed point for each continous function.

excitement when Brouwer first proved
the result over one hundred years ago.
Before we proceed, we must establish an auxiliary topological fact. For
this purpose that we are going to use
Poincaré's homotopy theory.

Lemma 5.3.2. *Let U, V be geometric*
figures and g : U → V a continuous
function. If γ is a closed curve in U
that can be continuously deformed to
a point, then g(γ) is a subset of V
that is also a closed curve that can be
continuously deformed to a point.

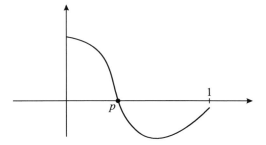

Figure 5.11.   The intermediate value theorem.

This statement makes good sense. Obviously a continuous function will not take a closed curve and open it up into a *non*-closed curve; that is antithetical to the notion of continuity. And if we imagine a flow of curves, beginning with γ, that merges to a point in U, then their images under g will be a flow of curves in V that merges to a point in V.

Definition 5.3.3. Let $\overline{D}$ be the closed unit disk (i.e., the unit disk together with its boundary) as shown in Figure 5.12. Let C denote the boundary circle of $\overline{D}$. A continuous function $h : \overline{D} \to C$ that fixes each point of C is called a *retraction* of $\overline{D}$ onto C. See Figure 5.13.

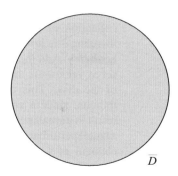

$\overline{D}$

Figure 5.12. The closed unit disk.

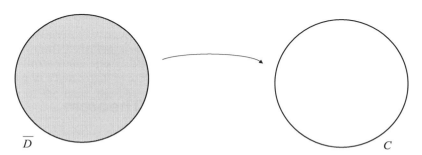

$\overline{D}$ C

Figure 5.13. A retraction.

Proposition 5.3.4. *There does not exist any retraction from $\overline{D}$ onto C.*

The proof of this result requires sophisticated ideas from homotopy theory—part of algebraic topology as developed by Poincaré over 100 years ago. We cannot provide the details here.

Here is Brouwer's famous theorem:

Theorem 5.3.5. *Let $\overline{D}$ be the closed unit disk. Let $f : \overline{D} \to \overline{D}$ be a continuous function. Then there is a point $P \in \overline{D}$ such that $f(P) = P$.*

Proof. At the risk of offending Brouwer himself, we provide a proof by contradiction. Suppose that there is such a map f that does *not* possess a fixed point. Then, for each point $x \in \overline{D}$, $f(x) \neq x$. But then we can use f to construct a retraction of $\overline{D}$ onto C as follows. Examine Figure 5.14. You can see that the segment that begins at $f(x)$, passes through x, and ends at a point $r(x)$ in C gives us a mapping

$$x \longmapsto r(x) \in C = \partial D.$$

Here ∂ is standard mathematical notation for "boundary."

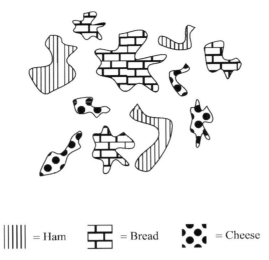

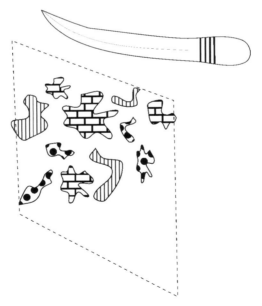

Figure 5.18.   A generalized ham sandwich.

See Figure 5.19. The proof, which is too complicated to present here, is a generalization of the intermediate value property that we used to prove the fixed-point theorem in dimension 1.

Figure 5.19.   The generalized ham sandwich theorem.

In fact it is worth pondering this matter a bit further. Let us consider the generalized ham-sandwich theorem in dimension 2. In this situation we cannot allow the generalized ham sandwich to have three ingredients. In fact, in dimension 2, the generalized ham sandwich has only bread and ham. No cheese. Then the same result is true: a single linear cut will bisect the ham and bisect the bread. Examine Figure 5.20 and convince yourself that with ham and cheese and bread configured as shown in dimension 2 there is no linear cut that will bisect all three

Figure 5.20. Limitations on the generalized ham sandwich theorem in dimension two.

quantities. But *any two* of the ham, bread, and cheese may be bisected by a single linear cut.

In dimension 4, we can add a fourth ingredient to the generalized ham sandwich—such as turkey. And then there is a single hyperplanar slice that will bisect each of the four quantities: turkey, ham, cheese, and bread. This is all pretty abstract, and we cannot discuss the details here (see [GAR] for a nice treatment).

5.5 Much Ado about Proofs by Contradiction

As we have seen, L.E.J. Brouwer used a "proof by contradiction" to establish his celebrated fixed-point theorem. But he later repudiated this methodology, claiming that all mathematical existence theorems should be established constructively. Today in fact there *are* constructive methods for proving versions of the Brouwer fixed-point theorem (see [BIS1]). We shall discuss one such method—due to Isaac Newton—below.[54]

At first Brouwer was a lone wolf, preaching his doctrine of constructivism by himself. But, over time, he gained adherents. Theoretical computer scientists have an interest in constructivism, as a computer is but a mechanical device for carrying our mathematical operations in a constructive manner. Proofs by contradiction have their place in the theory of computers, yet the computer itself is a constructivist tool.

The world was somewhat taken by surprise when, in 1968, mathematician Errett Bishop (1928–1983) came out in favor of constructivism. He was another researcher who had made his reputation by giving a number of dazzling proofs using the method of proof by contradiction. But then he rejected the methodology. We shall say more about Bishop later.

It is perhaps worth examining the 1-dimensional version of the Brouwer fixed-point theorem to see whether we can think about it in a constructivist manner. Refer to Section 5.3 to review the key ideas of the proof. The main step is that we had a function $g(x) = f(x) - x$, and we had to find a place where the function vanishes. We knew that $g(0) > 0$ and $g(1) < 0$,

[54]For completeness, we must note that Newton's method predates Brouwer by a couple of hundred years!

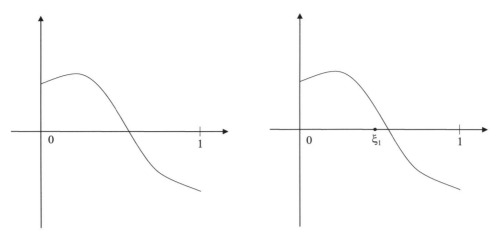

Figure 5.21. A smooth function as a candidate for Newton's method.

Figure 5.22. A first guess for Newton's method.

so we could invoke the intermediate value property of continuous functions to conclude that there is a point ξ between 0 and 1 such that $g(\xi) = 0$. It follows that $f(\xi) - \xi = 0$ or $f(\xi) = \xi$.

All well and good. This is a plausible and compelling proof. But it is not constructive. The existence of ξ is just that: an abstract existence proof. In general we cannot say what ξ is, or how to find it. In case g is a *smooth* function—i.e., no corners on its graph—then there is a technique of Isaac Newton that often gives a constructive method for finding ξ.[55]

We describe it briefly now.

So imagine our smooth function g that is positive at 0 and negative at 1—see Figure 5.21. In order to invoke Newton's method, we must start with a first guess ξ_1 for ξ. See Figure 5.22. Now the idea is to take the tangent line[56] to the graph at the point corresponding to ξ_1—see Figure 5.23—and see where *it* intersects the x-axis. This will be our second approximation ξ_2 to the number ξ that we seek. Figure 5.24 shows how the second guess is a considerable improvement over the first guess. In fact, calculations show that one usually doubles the number of decimal places of accuracy with each iteration of Newton's method.

One may apply this argument once again—taking a tangent line to the graph at ξ_2 and seeing where it intersects the x-axis. This will result in another notable improvement (called ξ_3) to our guess. Iterating Newton's method, one usually obtains a rapidly converging sequence of approximations to the true root (or zero) of the function.

---

[55]We should stress that there are examples in which Newton's method does *not* give the desired result—i.e., find the fixed point. Imposing smoothness does not address the fundamental epistemological issue raised by L.E.J. Brouwer—one still cannot in general find a fixed point constructively. But, as a practical matter, Newton's method usually works, and it gives a constructive procedure for producing a sequence that will converge to a fixed point. Newton's method is important, and is the basis for a big area of mathematics these days that is known as *numerical analysis*.

[56]This tangent line can be explicitly calculated using methods of calculus.

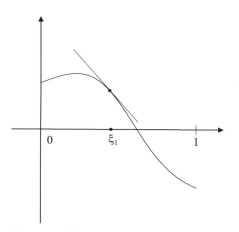

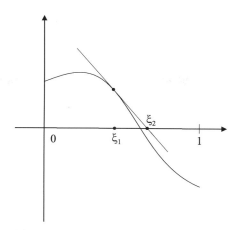

Figure 5.23. The second approximation for Newton's method.

Figure 5.24. The second guess is an improvement over the first guess.

As a simple illustration of Newton's method, suppose we wish to calculate $\sqrt{2}$—the famous number that Pythagoras proved to be irrational. This number is a root (or zero) of the polynomial equation $p(x) = x^2 - 2 = 0$. Let us take our initial guess to be $\xi_1 = 1.5$. We see that $1.5^2 = 2.25$, so this guess is a very rough approximation to the true value that we seek. It turns out that if one carries out the calculations needed (as we described above) for Newton's method—and these are quite straightforward if one knows a little calculus—then

$$\xi_{n+1} = \frac{\xi_n}{2} + \frac{1}{\xi_n}.$$

Applying this simple formula with $n = 1$, we find that

$$\xi_2 = \frac{\xi_1}{2} + \frac{1}{\xi_1}.$$

Now plugging in the value of our initial guess $\xi_1 = 1.5$, we find that

$$\xi_2 = 1.416666\ldots. \hspace{2cm} (*)$$

Since the true value of $\sqrt{2}$, accurate to twelve decimal places, is $\sqrt{2} = 1.414213562373$, we find that just one iteration of Newton's method gives us $\sqrt{2}$ to two decimal places of accuracy (only one decimal place if we round it off).

Now let us apply Newton's method again. So

$$\xi_3 = \frac{\xi_2}{2} + \frac{1}{\xi_2}.$$

Plugging in the value of ξ_2 from $(*)$, we find that

$$\xi_3 = 1.414215\ldots. \tag{$\star$}$$

Now we have $\sqrt{2}$ to four decimal places of accuracy.

Let us apply Newton's method one more time. We know that

$$\xi_4 = \frac{\xi_3}{2} + \frac{1}{\xi_3}.$$

Plugging in the value of ξ_3 from $(\star)$ now yields

$$\xi_4 = 1.41421356237\ldots.$$

Thus, with three simple iterations of Newton's method, we have achieved eleven decimal places of accuracy. Notice that the degree of accuracy at least doubled with each application of the technique!

Newton's method is the most fundamental technique in that branch of mathematics known as *numerical analysis*. The idea of numerical analysis is to use approximation methods—and computer calculation—to obtain answers to otherwise intractable problems.[57]

The use of numerical analysis methods involves another paradigm shift, for we are no longer finding precise solutions. Instead, we specify a desired degree of accuracy (in terms of the number of decimal places, for example) and then we find an approximate solution that has that number of decimal places of accuracy. In practical situations, this sort of approximation is usually satisfactory. In ordinary carpentry, one rarely requires more than 1 or 2 decimal places of accuracy. In highly technical engineering applications, perhaps 5 or 6 decimal places of accuracy are needed. In the design of microchips we might need 11 or 12 decimal places of accuracy. But one rarely needs the exact solution (even though the exact solution is of theoretical interest, and is no doubt what the mathematician would like to see).

As we have mentioned, Newton's method typically doubles the number of decimal places of accuracy with each iteration of the method. So it is a very effective device when used in conjunction with a computer.

5.6   Errett Bishop and Constructive Analysis

Errett Bishop was one of the great *geniuses* of mathematical analysis in the 1950s and 1960s. He made his reputation by devising fiendishly clever proofs about the structure of spaces of functions. Many of his proofs were indirect proofs—that is, proofs by contradiction.

Bishop underwent some personal changes in the mid- to late-1960s. He was a professor of mathematics at U.C. Berkeley and he was considerably troubled by all the political unrest on campus. After a time, he felt that he could no longer work in that atmosphere.

---

[57]Here we performed the calculations for $\sqrt{2}$ by hand, and this just took a few minutes. But a computer could do the calculations in a fraction of a second.

Also Bishop's gifted mathematician sister Mary Bishop apparently committed suicide. So Bishop was undergoing a certain amount of personal *angst*. He arranged to transfer to U.C. San Diego. At roughly the same time, Bishop became convinced that proofs by contradiction were fraught with peril. He wrote a remarkable and rather poignant book [BIS1] which touts the philosophy of constructivism—similar in spirit to L. E. J. Brouwer's ideas from fifty years before. Unlike Brouwer, Bishop really put his money where his mouth was. In the pages of his book, Bishop is actually able to develop most of the key ideas of mathematical analysis without resorting to proofs by contradiction. Thus he created a new field of mathematics called "constructive analysis."

A quotation from Bishop's preface to his book gives an indication of how that author himself viewed what he was doing:

> Most mathematicians would find it hard to believe that there could be any serious controversy about the foundations of mathematics, any controversy whose outcome could significantly affect their own mathematical activity.

He goes on to say:

> It is no exaggeration to say that a straightforward realistic approach to mathematics has yet to be tried. It is time to make the attempt.

In a perhaps more puckish mood, Bishop elaborates:

> Mathematics belongs to man, not to God. We are not interested in properties of the positive integers that have no descriptive meaning for finite man. When a man proves a positive integer to exist, he should show how to find it. If God has mathematics of His own that needs to be done, let Him do it Himself.

But our favorite Errett Bishop quotation, and the one that bears most closely on the theme of this book, is this:

> A proof is any completely convincing argument.

Bishop's arguments in *Methods of Constructive Analysis* [BIS1] were, as was characteristic of Bishop, devilishly clever. The book had a definite impact, and caused people to reconsider the methodology of modern analysis. Bishop's acolyte and collaborator D. Bridges produced the revised and expanded version [BIB] of his work (published after Bishop's death), and there the ideas of constructivism are carried even further.

Bishop applied his constructive approach to several other areas of mathematics. In 1972 he published *Constructive Measure Theory* [BIS2]. In the following year Bishop published *Schizophrenia in Contemporary Mathematics* [BIC], which gave his principles of constructivism. He listed:

(A) Mathematics is common sense;
(B) Do not ask whether a statement is true until you know what it means;
(C) A proof is any completely convincing argument;
(D) Meaningful distinctions deserve to be preserved.

5.7 Nicolas Bourbaki

The turn of the twentieth century saw the dawn of the modern age of logic. Mathematicians realized that the intellectual structure of mathematics was rather chaotic. There were no universally accepted standards of rigor. Different people wrote up proofs of their theorems in different ways. Some rather prominent mathematicians rarely proved anything rigorously.

From our current perspective of over 100 years distance, it is not clear how much of the mathematical climate around 1900 is due to a lack of coherence in the subject and how much is a reflection of a large number of *geniuses* working to the limits of their capacity, and in relative intellectual isolation. In today's mathematical climate, any worker is instantly answerable (because of the Internet, and because of the worldwide, closely knit scholarly community) to mathematicians in Australia, Japan, France, Turkey, and other points around the globe. It is virtually impossible to work in isolation. For the most part, a mathematician who works less than rigorously, who does not follow the well-established rules of the game, is quickly sidelined.[58] That was not the state of the art at the turn of the twentieth century.

One hundred or so years ago there was also no universally accepted language of mathematics. Different technical terms meant different things to different people. The foundations of geometry in France looked different from the foundations of geometry in England, and those in turn looked different (at least in emphasis) from the foundations of geometry in Germany. America was a fifth wheel in the mathematics game. From the point of view of the mandarins at the great world centers in Paris and Berlin and Göttingen, there had never been a great American mathematician—which is to say that nobody in the United States had ever proved a great theorem—one that was recognized by the authorities in the great intellectual centers of Europe. Appreciation of American mathematics would come somewhat later, through the work of G.D. Birkhoff, Norbert Wiener, and others.

David Hilbert of Göttingen was considered one of the premier intellectual leaders of European mathematics. Just as an indication of his preeminence, he was asked to give the keynote address at the Second International Congress of Mathematicians which was held in Paris in 1900. What Hilbert did at that meeting was earthshaking—from a mathematical point of view. He formulated twenty-three problems[59] that he thought should serve as beacons in the mathematical work of the twentieth century. On the advice of Hurwitz and Minkowski, Hilbert abbreviated his remarks and presented only ten of these problems in his lecture. But soon thereafter a more complete version of Hilbert's ideas was published in several countries.

For example, in 1902 the *Bulletin of the American Mathematical Society* published an authorized translation by Mary Winston Newson[60] (1869–1959). This version described all twenty-three of the unsolved mathematics problems that Hilbert considered to be of the first

---

[58] Of course there are exceptions. Fractal geometers have forged their own path, and developed a version of mathematics that is largely phenomenological. Chaoticists, numerical analysts, low-dimensional topologists all do mathematics a bit differently.

[59] In point of fact, Hilbert originally had twenty four problems on his list. But at the last minute he omitted one on proof theory. It was later found in his notes.

[60] Newson was the first American woman to earn a Ph.D. degree at the university in Göttingen.

rank, and for which it was of the greatest importance to find a solution. In fact, Hilbert's 18th problem contains the Kepler sphere-packing problem, which we discuss in detail in Chapter 9. The first person to solve a Hilbert problem was Max Dehn in the year 1900. It was problem number 3, about cutting up one polyhedron into pieces that can be reassembled to form another.[61] Dehn was a professor at the University of Frankfurt, and also spent time at the Illinois Institute of Technology.

It is often said that Hilbert was the last mathematician to be conversant with all parts of mathematics. He wrote fundamental texts on all the basic components of the mathematics of the day. So it was with ease that Hilbert could, in his lecture to the International Congress in 1900, survey the entire subject of mathematics and pick out certain areas that demanded attention and offered particularly tantalizing problems. What are now known as the "Hilbert problems" cover algebra, geometry, differential equations, combinatorics, real analysis, logic, complex analysis, and many other parts of mathematics as well. Hilbert's message was that these twenty-three were the problems that mathematicians of the twentieth century should concentrate their efforts on solving.

Of course Hilbert's name carried considerable clout, and the mathematicians in attendance paid careful attention to his admonitions. They took the problems home with them and in turn disseminated them to their peers and colleagues. We have noted that Hilbert's remarks were written up and published, and thereby found their way to universities all over the world. It rapidly became a matter of great interest to solve a Hilbert problem, and considerable praise and *encomia* were showered on anyone who did so.

Today ten of the original Hilbert problems (3, 7, 10, 11, 13, 14, 17, 19, 20, 21) can be said to have been solved completely. Eight others (1, 2, 5, 9, 12, 15, 18, 22) have solutions that are perhaps not universally accepted. There are four (4, 6, 16, 23) that are so inchoate that it is not clear that there will ever be an identifiable "solution." And then there is number 8, the Riemann hypothesis, for which there is no solution in sight. The references [GRA] and [YAN] give a detailed historical accounting of the colorful saga of the Hilbert problems.

One of Hilbert's overriding passions was logic, and he wrote an important treatise in the subject [HIA]. Since Hilbert had a universal and comprehensive knowledge of mathematics, he thought carefully about how the different parts of the subject fit together. And he worried about the axiomatization of the subject. Hilbert believed fervently that there ought to be a universal (and rather small) set of axioms for mathematics, and that all mathematical theorems should be derivable from those axioms.[62] But Hilbert was also fully cognizant of the rather uneven history of mathematics. He knew all too well that much of the literature was riddled with errors, inaccuracies and inconsistencies.

Hilbert's geometry book [HIL], which may well be his most important work (though others may argue for his number theory, his functional analysis, or his invariant theory),

---

[61] The book [GRA] recounts that in fact a significant special case of this third Hilbert problem was solved even before Hilbert posed it!

[62] We now know, thanks to work of Kurt Gödel (see Section 0.10), that in fact Hilbert's dream cannot be fulfilled. At least not in the literal sense. But it is safe to say that most working mathematicians take Hilbert's program seriously, and most of us approach the subject with this ideal in mind.

corrected many errors in Euclid's *Elements* and refined Euclid's axioms to the form in which we know them today. And his logic book lays out his program for the twentieth century to formalize mathematics, to uniformize its language, and to put the subject on a solid analytical footing. To repeat, Hilbert's hopes were rather dashed by the bold and ingenious work of Gödel that was to come thirty years later. Russell's paradox (see Section 1.10) was already making people mighty nervous. But the fact remains that Hilbert was an extremely prominent and influential scholar. People attended carefully to his teachings, and they adopted his program enthusiastically. Thus David Hilbert had an enormous influence over the directions that mathematics was taking at the beginning of the twentieth century.

There had long been a friendly rivalry between French mathematics and German mathematics. Although united by a common subject everyone loved, these two groups practiced mathematics with different styles, different emphases, and different priorities. The French took Hilbert's program for mathematical rigor very seriously, but it was in their nature then to endeavor to create their own home-grown program. This project was ultimately initiated and carried out by a remarkable figure in the history of modern mathematics. His name is Nicolas Bourbaki.

Jean Dieudonné, the *great raconteur* of twentieth century French mathematics, tells of a custom at the École Normale Supérieure in France to subject first-year students in mathematics to a rather bizarre rite of initiation. A senior student at the university would be disguised as an important visitor from abroad; he would give an elaborate and rather pompous lecture in which several "well-known" theorems were cited and proved. Each of the theorems would bear the name of a famous or sometimes not-so-famous French general, and each was wrong in some very subtle and clever way. The object of this farce was for the first-year students to endeavor to spot the error in each theorem, or perhaps not to spot the error but to provide some comic relief.

In the mid-1930s, a cabal of French mathematicians—ones who were trained at the *notorious* École Normale—was formed with the purpose of writing definitive texts in the basic subject areas of mathematics. They ultimately decided to publish their books under the *nom de plume* Nicolas Bourbaki. In fact the inspiration for their name was an obscure French general named Charles Denis Sauter Bourbaki. This general, so it is told, was once offered the chance to be King of Greece, but (for unknown reasons) he declined the honor. Later, after suffering an embarrassing retreat in the Franco-Prussian War, Bourbaki tried to shoot himself in the head—but he missed. Bourbaki's name had been used in the tomfoolery at the École Normale. Bourbaki was quite the buffoon. When the young mathematicians André Weil (1906–1998), Jean Delsarte (1903–1968), Jean Dieudonné (1906–1992), Jean Coulomb (1904–1999), Charles Ehresmann (1905–1979), Szolem Mandelbrot (1899–1983), René de Possel (1905–1974), Claude Chevalley (1909–1984), and Henri Cartan (1904–2008)[63] decided to form a secret organization (named the Association des Collaborateurs de Nicolas Bourbaki) that was dedicated to writing definitive texts in the basic subject areas of mathematics, they determined to name themselves after someone completely ludicrous. For what

---

[63] Jean Leray (1906–1998) and Paul Dubreil (1904–1994) were initially part of the group, but they both dropped out before the work actually got underway.

they were doing was of *utmost* importance for their subject. So it seemed to make sense to give their work a thoroughly ridiculous byline.

Thus, through a sequence of accidents and coincidences, it came about that some of the former students of the École Normale Supérieure banded together and decided to assemble an ongoing work that would systematically describe and develop all the key ideas in modern mathematics. Today it can safely be said that "Nicolas Bourbaki" is one of the most famous and celebrated mathematical names of modern times. He is the author of an extensive, powerful, and influential series of mathematics books. But "Nicolas Bourbaki" is actually an *allonym* for an anonymous group of distinguished French mathematicians.

There are other, not necessarily contradictory, stories of how Bourbaki came about. André Weil tells that in 1934 he and Henri Cartan were constantly squabbling about how best to teach Stokes's theorem in their respective courses at Strasbourg. He says,

> One winter day toward the end of 1934, I thought of a brilliant way of putting an end to my friend's [Cartan's] persistent questioning. We had several friends who were responsible for teaching the same topics in various universities. "Why don't we get together and settle such matters once and for all, and you won't plague me with your questions any more?" Little did I know that at that moment Bourbaki was born.

Weil claims that the name "Bourbaki" has an even longer history. In the early 1920s, when they were students at the École Normale, one of the older students (named Raoul Husson) donned a false beard and a strange accent and gave a much-advertised talk under the *nom de plume* of a fictitious Scandinavian. The talk was balderdash from start to finish, and concluded with a "Bourbaki theorem," which left the audience speechless. One of the École's students claimed afterward to have understood every word.

Jean Dieudonné describes the philosophy for what is proper grist for the Bourbaki books as follows:

> ... those which Bourbaki proposes to set forth are generally mathematical theories almost completely worn out already, at least in their foundations. This is only a question of foundations, not details. These theories have arrived at the point where they can be outlined in an entirely rational way. It is certain that group theory (and still more analytical number theory) is just a succession of contrivances, each one more extraordinary than the last, and thus extremely anti-Bourbaki. I repeat, this absolutely does not mean that it is to be looked down upon. On the contrary, a mathematician's work is shown in what he is capable of inventing, even new stratagems. You know the old story—the first time it is a stratagem, the third time a method. Well, I believe that greater merit comes to the man who invents the stratagem for the first time than to the man who realizes after three or four times that he can make a method from it.

As we have noted, the founding mathematicians in this new Bourbaki group came from the tradition of the École Normale Supérieure. This is one of the most elite universities in France, but it also has a long-standing tradition of practical joking. Weil himself tells of one particularly delightful episode. In 1916, Paul Painlevé (1863–1933) was a young and brilliant professor at the Sorbonne. He was also an examiner for admission to the École

Normale Supérieure. Each candidate for admission had to undergo a rigorous oral exam, and Painlevé was on the committee. So the candidates came early in the morning and stood around the hall outside the examination room awaiting their turn. On one particular day, some of the more advanced students of the École began to chat with the novices. They told the youngsters about the fine tradition of practical joking at the school. They said that one of the standard hoaxes was that some student would impersonate an examiner, and then ridicule and humiliate the student being examined. The students should be forewarned.

Armed with this information, one of the students went in to take the exam. He sat down before the extremely youthful-looking Painlevé and blurted out, "You can't put this over on me!" Painlevé, bewildered, replied, "What do you mean? What are you talking about?" So the candidate smirked and said, "Oh, I know the whole story, I understand the joke perfectly, you are an impostor." The student sat back with his arms folded and waited for a reply. And Painlevé said, "I'm Professor Painlevé, I'm the examiner . . . ."

Things went from bad to worse. Painlevé insisted that he was a professor, but the student would not back down. Finally, Painlevé had to go ask the Director of the École Normale to come in and vouch for him.

When André Weil used to tell this story, he would virtually collapse in hysterics.

In later years, Weil was at a meeting in India and told his friend Kosambi the story of the incidents that led to the formation of Bourbaki. Kosambi then used the name "Bourbaki" in a parody that he passed off as a contribution to the proceedings of some provincial academy. On the strength of this development, the still-nascent Bourbaki group determined absolutely that this would be its name. Weil's wife Eveline became Bourbaki's godmother and baptized him Nicolas.

Weil concocted a biography of Bourbaki and alleged him to be of "Poldavian descent." Nicolas Bourbaki of Poldavia submitted a paper to the journal *Comptes Rendus* of the French Academy to establish his bona fides. Élie Cartan (1869–1951) and André Weil exercised considerable political skill in getting the man recognized and the paper accepted. It turns out that "Poldavia" was another concoction of the practical jokers at the École Normale. Puckish students frequently wrote letters and gave speeches on behalf of the beleaguered Poldavians. One such demagogue gave a speech that ended, "And thus I, the President of the Poldavian Parliament, live in exile, in such a state of poverty that I do not even own a pair of trousers." He climbed onto a chair and was seen to be in his undershorts.

In 1939, André Weil was living in Helsinki. On November 30 of that year, the Russians conducted the first bomb attack on Helsinki. Shortly after the incident, Weil was wandering around the wrong place at the wrong time; his squinty stare and obviously foreign attire brought him to the attention of the police, and he was arrested. A few days later the authorities conducted a search of Weil's apartment in his presence. They found

- Several rolls of stenotypewritten paper at the bottom of a closet; Weil claimed that these were the pages of a Balzac novel.
- A letter, in Russian, from Pontryagin. It was arranging a visit of Weil to Leningrad.
- A packet of calling cards belonging to Nicolas Bourbaki, member of the Royal Academy of Poldavia.
- Some copies of Bourbaki's daughter Betti's wedding invitations.

In all, this was an incriminating collection of evidence. Weil was slammed into prison for good.

A few days later, on December 3, Rolf Nevanlinna (1895–1980)—at that time a reserve colonel on the general staff of the Finnish Army—was dining with the chief of police. (It should be noted that Nevanlinna was a distinguished mathematician—a complex analyst—in his own right. He was teacher to future Fields Medalist Lars Ahlfors.) Over coffee, the chief allowed that, "Tomorrow we are executing a spy who claims to know you. Ordinarily I wouldn't have troubled you with such trivia, but since we're both here anyway, I'm glad to have the opportunity to consult you." "What is his name?" inquired Nevanlinna. "André Weil." You can imagine Nevanlinna's shock—for André Weil was a world-renowned mathematician of the first rank. But he maintained his composure and said, "I know him. Is it really necessary to execute him?" The police chief replied, "Well, what do you want us to do with him?" "Couldn't you just deport him?" asked Nevanlinna innocently. "Well, there's an idea; I hadn't thought of it," replied the chief of police. And so André Weil's fate was decided.

When Weil was deported from Finland he was taken in custody to England and then to France. In France he was a member of the army reserves, and he was immediately jailed for failure to report for duty. He liked to say many years later that jail was the perfect place to do mathematics: it was quiet, the food was not bad, and there were few interruptions. In fact Weil wrote one of his most famous works—the book *Basic Number Theory* [WEIL1]— during his time in prison. In later years Hermann Weyl (1885–1955) threatened to have Weil put back in jail so that he would be more productive! In fact Weil himself wrote to his wife from his jail cell with the following sentiment:

> My mathematics work is proceeding beyond my wildest hopes, and I am even a
> bit worried—if it is only in prison that I work so well, will I have to arrange to
> spend two or three months locked up every year?

Reading his memoir [WEIL2], one gets the sense that the young Weil felt that the Bourbaki group was in effect reinventing mathematics for the twentieth century. In particular, they endeavored to standardize much of the terminology and the notation. Weil in fact takes credit for inventing the notation $\emptyset$ for the empty set (i.e., the set with no elements). The way he tells it, the group was looking for a good way to denote this special set; Weil was the only person in the group who knew the Norwegian alphabet, and he suggested that they coopt this particular letter. In fact, this consideration went into the *first* Bourbaki book—on set theory! It is remarkable—at least to someone with mathematical training—that the notation for the empty set was still being debated in the late 1930s.

The Bourbaki group was formed in the 1930s. The first preliminary meeting was in late 1934, and the clan quickly developed the rigorous habit of having two one-week meetings and one two-week meeting every year. The first formal Bourbaki meeting was in July, 1935 at Besse-en-Chandesse. They at first thought that they would be able to draft the essentials of all of mathematics in about three years. In point of fact, their first complete chapter (of just one book, on set theory) took four years.

Each of the founding members of the organization was himself a prominent and accomplished mathematician. Each had a broad view of the subject, and a clear vision of what

Bourbaki was meant to be and what it set out to accomplish. While the original goal of the group was to write a definitive text in mathematical analysis, ultimately it was decided that all of mathematics required examination and exposition. Even though the books of Bourbaki became well known and widely used throughout the world, the identity of the members of Bourbaki was a closely guarded secret. Their meetings, and the venues of those meetings, were kept under wraps. The inner workings of the group were not leaked by anyone, and in fact much disinformation was disseminated in order to keep the world at bay.

Over time, new members were added to Bourbaki—for instance Alexandre Grothendieck of the Institut des Hautes Études Scientifiques, Jean-Pierre Serre of the Collège de France, Samuel Eilenberg of Columbia University, Armand Borel of the Institute for Advanced Study, and Serge Lang of Columbia and Yale worked with Bourbaki. Others dropped out. But the founding group consisted exclusively of French mathematicians having a coherent vision for the needs of twentieth century mathematics. The Bourbaki group had few formal rules, but one of them was that a member must retire by the age of fifty.

The membership of Bourbaki was dedicated to the writing of the fundamental texts—in all the basic subject areas—in modern mathematics. Bourbaki's method for producing a book was as follows:

- The first rule of Bourbaki is that they would not write about a mathematical subject unless (i) it was basic material that any mathematics graduate student should know and (ii) it was mathematically "dead." This second desideratum meant that the subject area must no longer be an active area of current research in mathematics. Considerable discussion was required among the Bourbaki group to determine which were the proper topics for the Bourbaki books.
- Next there would be extensive and prolonged discussion of the chosen subject area: what are the important components of this subject, how do they fit together, what are the milestone results, and so forth. If there were several different ways to approach the subject (and often in mathematics this will be the case), then due consideration was given to which approach the Bourbaki book would take. The discussions we are describing here often took several days. The meetings were punctuated by long and sumptuous meals at good French restaurants.
- Once Bourbaki had agreed on a topic for a chapter, the job of writing the first draft could be taken up by any member who wanted it. This was a protracted affair, and could take a good many months. Jean Dieudonné, one of the founding members of Bourbaki, was famous for his skill and fluidity at writing. Of all the members of Bourbaki, he was perhaps the one who served most frequently as the scribe. Dieudonné was also a prolific mathematician and writer in his own right.[64]
- After a first draft had been written, copies would be made for the members of the Bourbaki group. At the next Bourbaki congress (so the meetings were dubbed), the new text would be read line-by-line. The group had to approve each line unanimously; any part of any Bourbaki book could be squelched by just a single member. The Bourbaki

---

[64]There is one particular incident that serves to delineate the two lines of his work. Dieudonné once published a paper under Bourbaki's name, and it turned out that this paper had a mistake in it. Some time later, a paper was published entitled, "On an error of M. Bourbaki," and the signed author was Jean Dieudonné.

members would read every word of every proof—assiduously and critically. Then the group would continue to hold their thrice-yearly congresses—punctuated as usual by sumptuous repasts at elegant French restaurants—in which they would go through the book line-by-line. The members of Bourbaki were good friends, and had the highest regard for each other as scholars, but they would argue vehemently over particular words or particular sentences in the Bourbaki text. They often said explicitly that confrontation led to productive work much more readily than congeniality. It would take some time for the group to work together through the entire first draft of a future Bourbaki book.

- After the group got through that first draft, and amassed a copious collection of corrections, revisions and edits, then a second draft would be created. This task could be performed by the original author of the first draft, or by a different author. And then the entire cycle of work would repeat itself.

It would take several years, and many drafts, for a new Bourbaki book to be created. The first Bourbaki book, on set theory, was published in 1939; Bourbaki books, and new editions thereof, have appeared as recently as 2005. So far there are volumes on 13 subject areas in the monumental series *Éléments de Mathématique*. But many of these subject areas are divided into "chapters," making it difficult to actually count how many books Bourbaki has penned. These books compose a substantial (though not complete) library of modern mathematics at the level of a first- or second-year graduate student. Topics covered range from abstract algebra to point-set topology to Lie groups to real analysis.

The writing in the Bourbaki books is crisp, clean, and precise. Bourbaki has a very strict notion of mathematical rigor. For example, *Bourbaki books, as a general rule, contain no pictures!* That's right. Bourbaki felt that pictures are an intuitive device, and have no place in a proper mathematics text. If the mathematics is written correctly, then the ideas should be clear—at least after sufficient cogitation. The Bourbaki books are written in a strictly logical fashion, beginning with definitions and axioms, and then proceeding with lemmas, propositions, theorems, and corollaries. Everything is proved rigorously and precisely. There are few examples and little explanation, mostly just theorems and proofs. There are no "proofs omitted," no "sketches of proofs," and no "exercises left for the reader." In every instance, Bourbaki gives us the whole enchilada.

The Bourbaki books have had a considerable influence in modern mathematics. For many years, other textbook writers sought to mimic the Bourbaki style. Walter Rudin was among them, and he wrote a number of influential texts without pictures and adhering to a strict formalism. In the 1950s, 1960s and 1970s, Bourbaki ruled the roost. This group of dedicated French mathematicians with the fictitious name had set a standard to which many aspired. It can safely be said that an entire generation of mathematics texts danced to the tune that was set by Bourbaki.

But fashions change. It is now a commonly held belief in France that Bourbaki caused considerable damage to the French mathematics enterprise. How could this be? Given the value system for mathematics that we have been describing in this book, given the passion for rigor and logic that is part and parcel of the subject, it would seem that Bourbaki would be our hero for some time to come. But no. There are other forces at play.

In some sense the Bourbaki books were the right books at the right time. In the 1940s, 1950s, and 1960s, there were few texts on advanced mathematics. So the Bourbaki texts really filled a void. Now there are many texts, and many different voices and perspectives to be heard. Also there are many avenues of mathematics that Bourbaki never explored.

One feature of Bourbaki is that the books were only about *pure* mathematics. There are no Bourbaki books about applied partial differential equations, or control theory, or systems science, or theoretical computer science, or cryptography, or graph theory, or any of the other myriad areas where mathematics is applied. There are also areas of pure mathematics that Bourbaki never touched—category theory is one, and number theory is another. Bourbaki has very little to say about logic. Bourbaki approaches all subjects—even analysis and geometry—from a very abstract and algebraic point of view.

Another feature of Bourbaki is that it rejects intuition of any kind.[65] Bourbaki books tend *not* to contain explanations, examples, or heuristics. One of the main messages of the present book is that we *record* mathematics for posterity in a strictly rigorous, axiomatic fashion. This is the mathematician's version of the reproducible experiment with control used by physicists and biologists and chemists. But we *learn* mathematics, we *discover* mathematics, we *create* mathematics using intuition and trial and error. We draw pictures. Certainly we try things and twist things around and bend things to try to make them work. Unfortunately, Bourbaki does not teach any part of this latter process.

Thus, even though Bourbaki has been a role model for what recorded mathematics ought to be, even though it is a shining exemplar of rigor and the axiomatic method, it is not necessarily a good and effective teaching tool. So, in the end, Bourbaki has not necessarily completed its grand educational mission. Whereas in the 1960s and 1970s it was quite common for Bourbaki books to be used as texts in courses all over the world, now the Bourbaki books are rarely used anywhere in classes. They are still useful references and helpful for self-study. But, generally speaking, there are much better texts written by other authors. We cannot avoid saying, however, that those "other authors" learned from Bourbaki. Bourbaki's influence is still considerable.

One amusing legacy that Bourbaki has left us is a special symbol used to denote a tricky passage in one of the Bourbaki texts. The reader who has traveled to Europe may remember the universal road signs to indicate a curvy or dangerous road. See Figure 5.25. This is the sign that Bourbaki uses—in the margin near a tricky passage—to denote mathematical danger. And this is perhaps the artifact of Bourbaki that lives on most universally. Many another textbook writer uses the "Bourbaki dangerous bend" to mark challenging passages in his/her text.

David Hilbert (see Section 5.1) and Nicolas Bourbaki have played an important role in the development of twentieth century mathematics. It can safely be said that today mathematics is practiced in the same way all over the world. Everyone uses the same terminology. Everyone embraces the same standards of rigor. Everyone uses the axiomatic

---

[65] In this sense Bourbaki follows a grand tradition. The master mathematician Carl Friedrich Gauss used to boast that an architect did not leave up the scaffolding so that people could see how he constructed a building. Just so, a mathematician does not leave clues as to how he constructed or found a proof.

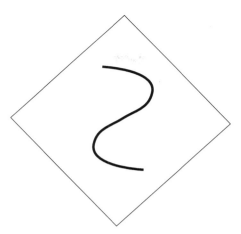

Figure 5.25. The Bourbaki "dangerous bend" sign.

method in the same way. And we can thank David Hilbert and Nicolas Bourbaki for showing us the way and setting the lasting example.

Jean Dieudonné (1906–1992) waxes euphoric in describing how Bourbaki cuts to the quick of mathematics:

> What remains: the archiclassic structures (I don't speak of sets, of course), linear and multilinear algebra, a little general topology (the least possible), a little topological vector spaces (as little as possible), homological algebra, commutative algebra, non-commutative algebra, Lie groups, integration, differentiable manifolds, Riemannian geometry, differential topology, harmonic analysis and its prolongations, ordinary and partial differential equations, group representations in general, and in its widest sense, analytic geometry. (Here of course I mean in the sense of Serre, the only tolerable sense. It is absolutely intolerable to use *analytic geometry* for linear algebra with coordinates, still called analytical geometry in the elementary books. Analytical geometry in this sense has never existed. There are only people who do linear algebra badly, by taking coordinates and this they call analytical geometry. Out with them! Everyone knows that analytical geometry is the theory of analytical spaces, one of the deepest and most difficult theories of all mathematics.) Algebraic geometry, its twin sister, is also included, and finally the theory of algebraic numbers.

When Ralph Boas was the Executive Editor of the *Mathematical Reviews*, he made the mistake of printing the opinion that Bourbaki does not actually exist—in an article for the *Encyclopædia Britannica*, no less. The *Encyclopædia* subsequently received a scalding letter, signed by Nicolas Bourbaki, in which he declared that he would not tolerate the notion that anyone might question his right to exist. To avenge himself on Boas, Bourbaki began to circulate the rumor that Ralph Boas did not exist. In fact, Bourbaki claimed that "Boas" was actually an acronym; the letters B.O.A.S. were actually a pseudonym for a group of the *Mathematical Reviews'* editors.

"Now, really, these French are going too far. They have already given us a dozen independent proofs that Nicolas Bourbaki is a flesh and blood human being. He writes papers, sends telegrams, has birthdays, suffers from colds, sends greetings. And now they want us to take part in their canard. They want him to become a member of the American Mathematical Society (AMS). My answer is 'No'." This was the reaction of J.R. Kline, secretary of the AMS, to an application from the legendary Nicolas Bourbaki.

5.8   Srinivasa Ramanujan and a New View of Proof

Srinivasa Ramanujan (1887–1920) was one of the most remarkable and talented mathematicians of the twentieth century. Born in poverty in the small village of Erode, India, Ramanujan attended a number of elementary schools. He showed considerable talent in all his subjects. At the age of thirteen Ramanujan began to conduct his own independent mathematical investigations. He determined how to sum various kinds of series (arithmetical and geometric). At age fifteen Ramanujan was shown how to solve a cubic, or third-degree, polynomial equation. He used that idea to devise his own method for solving the quartic, or fourth-degree, equation. His efforts to solve the quintic, or fifth degree, equation were doomed to failure, because Abel had shown many years before that this was impossible.

By age 17, Ramanujan had studied an old textbook of G.S. Carr and developed considerably in mathematical sophistication. He began his own investigations of Euler's constant and of Bernoulli numbers.

Ramanujan earned a scholarship to attend the Government College in Kumbakonam. He began his studies at age 17 in 1904, but was soon dismissed because he spent all his time on mathematics and let his other subjects languish. He studied hypergeometric series and elliptic functions on his own.

In 1906 Ramanujan endeavored to pass the entrance exams to the University of Madras. But he could pass only the mathematical part of the test, so he failed.

During all this time Ramanujan's health was quite fragile. He suffered from smallpox when he was 2 years old and other recurrent ailments of a serious nature throughout his young adult life. In 1909 Ramanujan had an arranged marriage with a ten-year-old girl named S. Janaki Ammal.

Ramanujan was able to publish some of his ideas about elliptic functions and Bernoulli numbers in the *Journal of the Indian Mathematical Society*. He gained thereby a considerable reputation as a young mathematical genius.

In order to support himself while he studied mathematics, Ramanujan occupied various low-paying clerkships. In 1912 and 1913, Ramanujan wrote to a number of prominent British mathematicians seeking advice and help with his work. These included M.J.M. Hill, E.W. Hobson, H.F. Baker, and G.H. Hardy. It was Hardy who immediately apprehended what a remarkable person this was, and who truly understood Ramanujan's genius; he decided to take action. This was surely the turning point of Ramanujan's life and of his mathematical career.

In 1914 Hardy was able to bring Ramanujan to Trinity College at Cambridge University. Ramanujan was an orthodox Brahmin and also a strict vegetarian. The former attribute made his travel very difficult, and the latter made it complicated for Ramanujan (a man of ill health) to find proper nourishment in World War I Britain. Nonetheless, Hardy and Ramanujan immediately began a most fruitful and dynamic mathematical collaboration.

But there were difficulties too, and that is really the point of the present discussion. Ramanujan had very little formal education. He did not see or understand the importance of mathematical proof. He just "saw" things in an almost mystical fashion, and let others worry about how to prove them. Often Ramanujan was right in profound and important ways. But he made mistakes too. His view was that the gods spoke to him and gave him these mathematical insights. Hardy was one of the most accomplished and powerful mathematicians of his day, and he found Ramanujan's view frustrating. Hardy himself could sometimes provide the needed proofs; but sometimes not.

Ramanujan found his ill health to be too much of a burden by 1919, and he returned to India. He died within a year of medical complications that may have been related to diet or vitamin deficiency (although some theorize that it was tuberculosis).

Ramanujan left behind a number of remarkable notebooks filled with his formulas and insights. But no proofs. Mathematicians even today are hard at work filling in the details so that we may understand the derivations of Ramanujan's astonishing relationships among numbers. We close with a famous story that illustrates Ramanujan's almost magical powers.

During Ramanujan's final days in Great Britain, he lay ill in a hospital. His friend and mentor Hardy went to visit him frequently. The great professor was not much of a conversationalist, and he struggled to find things to say to his friend. One day Hardy walked in the door and mumbled, "I noticed that the number of my taxicab was 1729. It seemed rather a dull number." Ramanujan's instantaneous reaction was, "No, Hardy! No, Hardy! It is a very interesting number. It is the smallest number expressible as the sum of two cubes in two different ways."

Once Hardy knew this fact he could prove it without much difficulty. But how did Ramanujan see it?

5.9 The Legend of Paul Erdős

Paul Erdős (1913–1996) was a very unusual man. Born to a middle class family in Hungary, Erdős was doted on by his parents. He did not butter his own bread until he was twenty one years old. His mother did everything for him—in part out of deference to his obviously remarkable and rare mathematical talents.

Erdős grew up to be a very powerful and creative mathematician. But he never held a regular job. He would travel from math department to math department, clutching his few possessions in a small flight bag, and collaborate with mathematicians whom he knew wherever he was. Of course, Erdős expected people to take care of him. He expected to be fed and housed, and for someone to wash his clothes and pay his expenses. Erdős had

He also wrote a famous article entitled

Applied mathematics is bad mathematics.

He is quick to say in the first paragraph of that article that he didn't really mean what the title says. But it is and was clear that the point was made.

One notable feature of Paul Halmos's career is that he held many different positions: University of Illinois, Institute for Advanced Study, Syracuse University, University of Chicago, University of Michigan, University of Hawaii, University of California at Santa Barbara, Indiana University, and University of Santa Clara. Halmos was alway looking to see what was around the next turn in the road; for him life, and especially mathematical life, was an adventure.

Halmos is particularly remembered for his memoir *I Want to Be a Mathematician: An Automathography*. Published in 1985, this is a particularly candid accounting not only of Halmos's life, but also of the many people and places with which he had interacted. Halmos pulls no punches in making frank remarks about living mathematicians, and how they had treated (or mistreated) Halmos; people reacted quite strongly to the book. Nonetheless, *I Want to Be a Mathematician* will be remembered as an important broadside about the mathematical life and what it means.

Halmos was an energetic mathematician, editor, writer, raconteur, and particularly a teacher. He had many Ph.D. students, and is remembered as a singularly outstanding classroom teacher and lecturer. Halmos had a profound influence on everyone who knew him. Like Erdős, he was the purest of pure mathematicians. Halmos stated theorems, and he proved them. Nothing else mattered.

5.11 Perplexities and Paradoxes

One upshot of the investigations into the foundations of mathematics that are the hallmark of twentieth century research is a variety of rather troubling paradoxes that have come to the fore. We have already mentioned Russell's paradox in Section 1.10. We take the opportunity in this section to indicate a few others.

The point of these paradoxes is that they are results that can be proved logically—and *correctly*—from the axioms of set theory as we know them. But these paradoxes are so counterintuitive, and so astonishing, that we cannot help but wonder whether mathematics is fraught with internal contradictions, or whether we are all on a fool's errand.

We modern mathematicians know better than this. All the paradoxes that are discussed here *can* be explained, and we shall give an indication of what those explanations are. It takes some time, and some effort, to become inured to these paradoxes. We may consider this treatment to be your first exposure.

5.11.1 Bertrand's Paradox

This paradox was discovered a few hundred years ago. It is part of the reason that the subject of probability theory had such a rocky start. The proper resolution of this paradox, and why it really harbors no inherent contradiction, did not come about until the late 1930s. So it is suitable grist for our mill here.

Fix a circle of radius 1. Draw the inscribed equilateral triangle as shown in Figure 5.26. We let ℓ denote the length of a side of this triangle. Suppose that a chord d (with length m) of the circle is chosen "at random."[66] What is the probability that the length m of d exceeds the length ℓ of a side of the inscribed triangle?

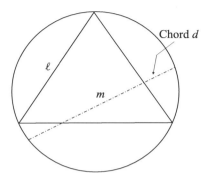

Figure 5.26. Bertrand's paradox.

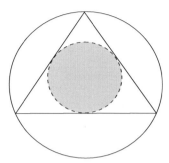

Figure 5.27. First look at Bertrand's paradox.

The "paradox" is that this problem has three different but equally valid solutions. We now present these apparently contradictory solutions in sequence. At the end we shall explain why it is possible for a problem like this to have three distinct solutions.

Solution 1. Examine Figure 5.27. It shows a shaded, open disk whose boundary circle is internally tangent to the inscribed equilateral triangle. If the center of the random chord d lies *inside* that shaded disk, then $m > \ell$. If the center of the random chord d lies *outside* that shaded disk, then $m \leq \ell$. Thus the probability that the length d is greater than the length ℓ is

$$\frac{\text{area of shaded disk}}{\text{area of unit disk}}.$$

But an analysis of the equilateral triangle (Figure 5.28) shows that the shaded disk has radius $1/2$ hence area $\pi/4$. The larger unit disk has area π. The ratio of these areas is $1/4$. We conclude that the probability that the length of the randomly chosen chord exceeds ℓ is $1/4$.

Solution 2. Examine Figure 5.29. We may as well assume that our randomly chosen chord is horizontal (the equilateral triangle and the chord can both be rotated so that the chord is horizontal and one side of the triangle is horizontal). Notice that if the height, from the base

---

[66]Without getting too formalistic, we can say that "at random" here means that you close your eyes and just grab a chord from the infinitely many that there are.

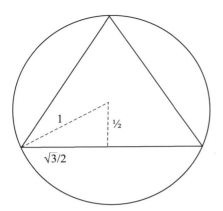

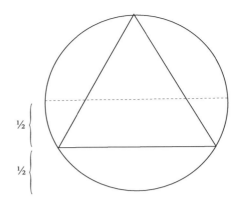

Figure 5.28. First look at Bertrand's paradox examined further.

Figure 5.29. Second look at Bertrand's paradox.

of the triangle, of the chord d is less than or equal to $1/2$ then $m \leq \ell$, while if the height is greater than $1/2$ (and not more than 1) then $m > \ell$. We thus see that there is probability $1/2$ that the length m of d exceeds the length ℓ of a side of the equilateral triangle.

Solution 3. Examine Figure 5.30. We may as well assume that one vertex of our randomly chosen chord occurs at the lower left vertex A of the inscribed triangle (by rotating the triangle we may always arrange this to be the case). Now look at the angle θ that the chord subtends with the tangent line to the circle at the vertex A (shown in Figure 5.31). If that angle is between $0°$ and $60°$ inclusive then the chord is shorter than or equal to ℓ. If the

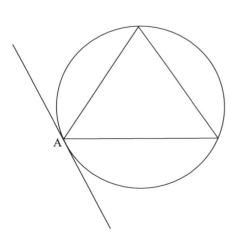

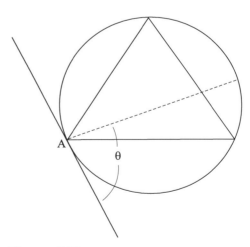

Figure 5.30. Third look at Bertrand's paradox.

Figure 5.31. More on the third look at Bertrand's paradox.

angle is strictly between $60°$ and $120°$ then the chord is longer than ℓ. Finally, if the angle is between $120°$ and $180°$ inclusive then the chord is shorter than ℓ. In sum we see that the probability is $60/180 = 1/3$ that the randomly chosen chord has length exceeding ℓ.

We have seen then three solutions to our problem. And they are different: we have found valid answers to be $1/4$, $1/2$, and $1/3$. How can a perfectly reasonable problem have three distinct solutions? And be assured that each of these solutions is correct! The answer is that when one is dealing with a probability space having infinitely many elements (that is, a problem in which there are infinitely many outcomes—in this case there are infinitely many positions for the random chord), then there are infinitely many different ways to fairly assign probabilities to those different outcomes. Our three distinct solutions arise from three distinct ways to assign probabilities: notice that one of these is based on area, one is based on height, and one is based on angle.

For many years, because of paradoxes such as this one, the subject of probability theory was in ill repute. It was not until the invention of a branch of mathematics called "measure theory" that the tools became available to put probability theory on a rigorous footing. The celebrated Soviet mathematician A. Kolmogorov is credited with this development. These ideas are treated in advanced courses on measure theory and probability.

5.11.2   The Banach–Tarski Paradox

Alfred Tarski (1902–1983) was one of the pioneering logicians of the twentieth century. Stefan Banach (1892–1945) was one of the great mathematical analysts. In 1924 they published a joint paper presenting the remarkable paradox that we are about to describe— a byproduct of the intense examination of the axioms of set theory that was the hallmark of the first half of the twentieth century.

> **The Banach–Tarski paradox:** *It is possible to take a solid ball of radius* 1*, break it up into* 7 *pieces, and reassemble those pieces into two solid balls of radius* 1*. Refer to Figure 5.32.*

How could this be? The statement seems to contradict fundamental principles of physics, and to undermine the notion that we have any concept of volume or distance. After all, we are beginning with a ball of volume $4\pi/3$ and ending up with two balls having total volume $8\pi/3$.

And in fact the Banach–Tarski paradox has even more dramatic formulations. It is actually possible to take a solid ball of radius 1, break it up into finitely many pieces, and reassemble those pieces into a full-sized replica of the Empire State Building. See Figure 5.33.

What could possibly be the explanation for this conundrum? It all has to do with measure theory, a subject pioneered by Henri Lebesgue (1875–1941) in 1901. The goal of measure theory is to assign a "measure," or size, or volume to every set. It turns out—and this is a consequence of the axiom of choice, about which we shall say more below—that this goal is an impossible one. Any attempt to assign a measure to every set is doomed to failure. Thus there are certain sets, which we can identify explicitly by a concrete criterion, called

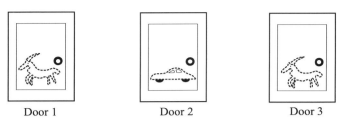

Door 1   Door 2   Door 3

Figure 5.34.   The Monte Hall problem.

Clearly the contestant will not pick the door that Monte Hall has already opened, since that has a goat behind it. So the issue is whether the contestant will switch from the currently selected door to the remaining door (the one that the contestant has not chosen and Monte Hall did not open). A naive approach would be to say there is an equal probability for there to be a goat behind the remaining door and behind the door that the contestant has already selected—after all, one door has a goat and one has a car. What is the point of switching? However, this naive approach does not take into account the fact that there are two distinct goats. A more careful analysis of cases occurs in our solution to the problem, and reveals a surprising answer.

Let us use a case-by-case analysis to solve the Monte Hall problem. We denote the goats by G_1 and G_2 (for goat one and goat two) and the car by C. For simplicity, we assume that the contestant will always select Door 3. We may not, however, assume that Monte Hall always reveals a goat behind Door 1; for there may not be a goat behind Door 1 (it could be behind Door 2). Thus there are several cases to consider:

Door 1	Door 2	Door 3
G_1	G_2	C
G_2	G_1	C
G_1	C	G_2
G_2	C	G_1
C	G_1	G_2
C	G_2	G_1

In general there are $n! = n \cdot (n-1) \cdot (n-2) \cdots 3 \cdot 2 \cdot 1$ different ways to arrange n objects in order. Thus there are $6 = 3!$ possible permutations of three objects. That is why there are six rows in the array. The table represents all the different ways that two goats and a car could be distributed behind the doors.

1. In the first case, Monte Hall will reveal a goat behind either Door 1 or Door 2. It is *not* to the contestant's advantage to switch (for the contestant has selected Door 3, which has the fancy car behind it), so we record **N**.

2. The second case is similar to the first, and it is not to the contestant's advantage to switch, and we record **N**.

3. In the third case, Monte Hall will reveal a goat behind Door 1, and it *is* to the contestant's advantage to switch. We record **Y**.

4. The fourth case is like the third, and it is to the contestant's advantage to switch. We record **Y**.

5. In the fifth case, Monte Hall will reveal a goat behind Door 2. It *is* to the contestant's advantage to switch, so we record **Y**.

6. The sixth case is like the fifth, it is to the contestant's advantage to switch, and we record **Y**.

Observe that the tally of our case-by-case analysis is four **Y**'s and just two **N**'s. Thus the odds are "two against one" in favor of switching after Monte Hall reveals the goat. Put in other words, the player has a 2/3 probability of improving his/her position for a win by switching doors.

The modern history of the Monte Hall problem is both amusing and alarming. As the reader may know, Marilyn vos Savant (1946–) is a popular newspaper columnist. (In point of fact "vos Savant" is not her real name—she was born Marilyn Mach. The surname "vos Savant" is borrowed from a favorite aunt.) She is reputed to have the highest IQ in history (although there is word that a child in China has a measurably higher IQ). Her syndicated column, *Ask Marilyn*, is premised on her high intellectual powers. She is very clever, and has a knack for answering difficult questions (she usually has the good sense to ask experts when confronted with a problem on which she has no expertise). She rarely makes a mistake, although there is a website called *Marilyn is Wrong!* [http://www.wiskit.com/marilyn.html], which claims to point out a number of her *faux pas*.

Actually Marilyn vos Savant was educated in St. Louis (my stomping ground). She attended Meramec Junior College, but did not graduate. She also attended Washington University (my institution), but did not graduate. She is married to Robert K. Jarvik (who invented the artificial heart), and makes her home in New York City.

Marilyn vos Savant gained particular celebrity when one of her readers (it is rumored to have been Steve Selvin) wrote in to ask about the "Monte Hall problem." She checked her facts and described in her column the correct solution, as we have discussed above. Woe is us!! Over 10,000 readers, at least 1,000 of these academic mathematicians (many writing on university letterhead!) wrote to Marilyn vos Savant and told her that she was in error. And some of them were not too polite about it. This was quite a debacle for American mathematics.

I have to say that this whole set of circumstances went to Marilyn vos Savant's head. After Andrew Wiles published his solution of Fermat's last theorem, Ms. Savant published a little book [SAV] claiming that his solution is incorrect. The basis for her daring allegation is twofold: (i) she offers a proof that the complex numbers do not exist (therefore nullifying Wiles's use of said number system) and (ii) she observes that Wiles uses hyperbolic geometry, in which the circle can be squared—but everyone knows that the circle *cannot be squared*— and this is a contradiction.

I wrote to Ms. vos Savant, and to her publisher, pointing out the error of her ways in the book [SAV]. And she answered! She said, "My mathematician friends and I had a good laugh over your submission. Keep those cards and letters coming." So much for scholarly discourse.

Martin Gardner is particularly chagrined for having been thanked for checking the math in vos Savant's book [SAV]. He asserts vehemently that he did not. Barry Mazur of Harvard was also asked to check the math, but he managed to dodge the bullet.

5.11.4 The Axiom of Choice

We alluded to the axiom of choice in our discussion of the Banach–Tarski paradox. Zermelo's axiom of choice (formally invented in 1904, but studied before that) has been one of the bugaboos of twentieth century mathematics.

In the following description of the axiom of choice, we shall use the notion of "subset." A set Y is said to be a subset of a set X if each element of Y is also an element of X. For example, let X be the positive integers and let Y be $\{2, 4, 6, 8\}$. Then Y is a subset of X. We write $Y \subset X$.

Simply stated, the axiom of choice says that if X is a set then there is a function that assigns to each subset S of X some element of S. Put in other words, if you have a collection of sets (all subsets of a given set), then you get to pick an element from each one.

This sounds harmless enough, but the axiom of choice has profound consequences. As we noted in Subsection 5.11.2, the axiom of choice implies the Banach–Tarski paradox. The axiom of choice is also equivalent to the statement that any set can be well ordered. This means that, if we are given a set X, then we can equip X with an ordering so that each subset of X has a least element. If the set X is the real numbers then this assertion is not at all obvious.

The axiom of choice is used extensively in algebra, logic, and category theory to perform needed constructions. The existence of maximal ideals, the existence of the algebraic closure, and many other essenetial parts of the theory depend on the axiom of choice. The axiom also comes up in analysis—for example in the proof of the Hahn–Banach theorem.

An old joke of Bertrand Russell says that if I have to choose one shoe from each of infinitely many pairs of shoes, I can do it easily. But if I have to choose one sock from each of infinitely many pairs of socks then I need the axiom of choice. Just what is old Bertie talking about? Well, with the shoes you can just say, "Pick the left shoe." But with socks you cannot. You need some mechanism for selecting a sock from each pair. The axiom of choice is such a mechanism.

6

The Tantalizing Four-Color Theorem

... the miracle of the appropriateness of the language of mathematics for the formulation of the laws of physics is a wonderful gift which we neither understand nor deserve. We should be grateful for it and hope that it will remain valid in the future and that it will extend ... to wide branches of learning.

—Eugene Wigner

The only proof capable of being given that an object is visible is that people actually see it ... In like manner, I apprehend, the sole evidence it is possible to produce that anything is desirable is that people do actually desire it.

—John Stuart Mill

Philosophers have frequently distinguished mathematics from the physical sciences. While the sciences were constrained to fit themselves via experimentation to the real world, mathematicians were allowed more or less free reign within the abstract world of the mind. This picture has served mathematicians well for the past few millennia, but the computer has begun to change this.

—J. Borwein, P. Borwein, R. Girgensohn, and S. Parnes

6.1 Humble Beginnings

In 1852 Francis W. Guthrie, a graduate of University College London, posed the following question to his brother Frederick:

> Imagine a geographic map of the earth (i.e., a sphere) consisting of countries only—no oceans, lakes, rivers, or other bodies of water. The only rule is that a country must be a single contiguous mass—in one piece, and with no holes [see Figure 6.1]. As cartographers, we wish to *color* the map so that no two adjacent countries will be of the same color [Figure 6.2—note that R, G, B, Y stand for red, green, blue, and yellow]. How many colors should the mapmaker keep in stock so that he can be sure he can color any map?

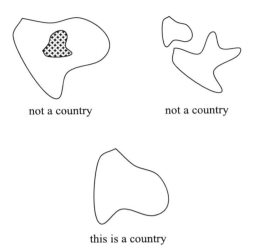

not a country   not a country

this is a country

Figure 6.1.   A country in the four-color problem.

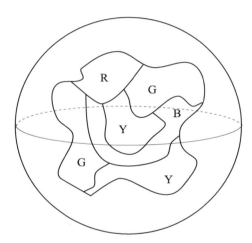

Figure 6.2.   No two adjacent countries are the same color.

Frederick Guthrie was a student of Augustus De Morgan (1806–1871), and ultimately communicated the problem to his mentor. The problem was passed around among academic mathematicians for a number of years (in fact De Morgan communicated the problem to William Rowan Hamilton (1805–1865)). The first allusion in print to the problem was by Arthur Cayley (1821–1895) in 1878.

The question became known as the *four-color problem*, because experimentation led everyone to believe that any map, as described by Guthrie, can be colored with four colors. Nobody could produce an example of a map that required five colors.

The mathematician Felix Klein (1849–1925) in Göttingen heard of the problem and declared that the only reason the problem had never been solved is that no capable mathematician had ever worked on it. *He*, Felix Klein, would offer a class, the culmination of which would be a solution of the problem. He failed.

In 1879, A. Kempe (1845–1922) published a solution of the four-color problem. That is, he showed that any map could be colored with four colors. Kempe's proof stood for 11 years. Then a mistake was discovered by P. Heawood (1861–1955). Heawood studied the problem further and came to a number of fascinating conclusions:

- Kempe's proof, particularly his device of "Kempe chains," *does* suffice to show that any map can be colored with five colors.
- Heawood showed that if the number of edges around each region in the map is divisible by 3, then the map is four-colorable.
- Heawood found a formula that gives an estimate for the "chromatic number" of practically any surface. Here the chromatic number $\chi(g)$ of a surface is the least number of colors it will take to color *any* map on that surface. In fact the formula is

$$\chi(g) \leq \left\lfloor \frac{1}{2} \left(7 + \sqrt{48g + 1}\right) \right\rfloor$$

so long as $g \geq 1$.

Here is how to read this formula. It is known, thanks to work of Camille Jordan (1838–1922) and August Möbius (1790–1868), that any surface in space is (equivalent to, or homeomorphic to) a sphere with handles attached. See Figure 6.3. The number of handles is called the *genus*, and we denote it by g. The Greek letter chi (χ) is the chromatic number of the surface—the least number of colors that it will take to color any map on the surface. Thus $\chi(g)$ is the number of colors that it will take to color any map on a surface that consists of the sphere with g handles. Next, the symbols $\lfloor \ \rfloor$ stand for the "greatest integer function." For example, $\lfloor \frac{9}{2} \rfloor = 4$ just because the greatest integer in the number "four and a half" is 4. Also $\lfloor \pi \rfloor = 3$ because $\pi = 3.14159\ldots$ and the greatest integer in the number pi is 3.

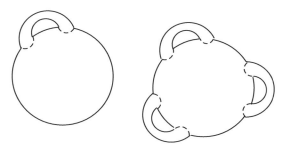

Figure 6.3. An embedded surface is a sphere with handles.

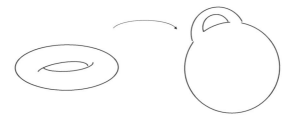

Figure 6.4.   The torus is a sphere with one handle.

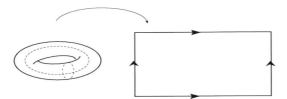

Figure 6.5.   Cutting the torus apart.

Now a sphere is a sphere with no handles, so $g = 0$. We may calculate that

$$\chi(g) \leq \left\lfloor \frac{1}{2}\left(7 + \sqrt{48 \cdot 0 + 1}\right) \right\rfloor = \left\lfloor \frac{1}{2}(8) \right\rfloor = 4 \,.$$

This is the four-color theorem! Unfortunately, Heawood's proof was valid only when the genus is at least 1. It gives no information about the sphere.

The torus (see Figure 6.4) is topologically equivalent to a sphere with one handle. Thus the torus has genus $g = 1$. Then Heawood's formula gives the estimate 7 for the chromatic number. And in fact we can give an example of a map on the torus that requires 7 colors—see below.

Here is what Figure 6.5 shows. It is convenient to take a pair of scissors and cut the torus apart. With one cut, the torus becomes a cylinder; with the second cut it becomes a rectangle. The arrows on the edges indicate that the left and right edges are to be identified (with the same orientation), and the upper and lower edges are to be identified (with the same orientation).

Now we color the torus by coloring the associated rectangle (as in Figure 6.5). We call our colors "1", "2", "3", "4", "5", "6", "7". The reader may verify that there are seven countries shown in our Figure 6.6, and every country is adjacent to every other (remember here that the upper and lower edges of the rectangle are identified or pasted together, and so are the left and right edges). Thus they all must have different colors! This is a map on the torus that requires 7 colors; it shows that Heawood's estimate is sharp for this surface.

Heawood was unable to decide whether the chromatic number of the sphere is 4 or 5. He was also unable to determine whether any of his estimates for the chromatic numbers

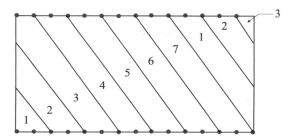

Figure 6.6.   A map on the torus that requires seven colors.

of various surfaces of genus $g > 1$ were sharp or accurate. That is, for the torus (the closed surface of genus 1), Heawood's formula says that the chromatic number does not exceed 7. Is that in fact the best number? Is there a map on the torus that really requires 7 colors? And for the torus with two handles (genus 2), Heawood's estimate gives a bound of 8. Is that the best number? Is there a map on the double torus that actually *requires* 8 colors? And so forth: we can ask the same question for every surface of every genus. Heawood could not answer these questions.

The mathematician P.G.Tait produced another resolution of the four-color problem in 1880. J.Petersen pointed out a gap in 1891. Another instance of 11 years lapsing before the error was found!

The four-color problem has a long and curious history. The great American mathematician G.D. Birkhoff did foundational work on the problem that allowed Philip Franklin (1898–1965) in 1922 to prove that the four-color conjecture is true for maps with at most 25 countries. H. Heesch made seminal contributions to the program, and in fact introduced the techniques of reducibility and discharging that were ultimately used by Appel and Haken in 1977 to solve the four-color problem. Walter Stromquist proved in his 1975 Harvard Ph.D. thesis [STR1] that, for any map with 100 or fewer countries, four colors will always suffice. See [STR2]. What is particularly baffling is that Ringel and Youngs were able to prove in 1970 that all of Heawood's estimates, for the chromatic number of any surface, are sharp. So the chromatic number of a torus is indeed 7. The chromatic number of a "super torus" with two holes is 8. And so forth. But the Ringel/Youngs proof does not apply to the sphere. They could not improve on Heawood's result that 5 colors will always suffice.

Then in 1974 there was blockbuster news. Using 1200 hours of computer time on the University of Illinois supercomputer, Kenneth Appel and Wolfgang Haken showed that in fact 4 colors will always work to color any map on the sphere. Their technique was to identify 633 fundamental configurations of maps (to which all others can be reduced) and to prove that each of them is reducible in the sense of Heesch. But the number of "fundamental examples" was very large, and the number of reductions required was beyond the ability of any human to count. And the logic is extremely intricate and complicated. Enter the computer.

draws together all the history and the techniques and gives a definitive description of the Appel–Haken effort. It also establishes that a map can be four-colored in polynomial time (see Section 11.7 for this concept).

In fact, for many years after that, the University of Illinois mathematics department had a postmark that appeared on every outgoing letter from their department. It read:

FOUR COLORS SUFFICE

Quite a triumph for Appel and Haken and their supercomputer.

But it seems as though there is always trouble in paradise. According to one authority, who prefers to remain nameless, errors continue to be discovered in the Appel–Haken proof. There is considerable confidence that any error that is found can be fixed. And invariably the errors *are* fixed. But the stream of errors never seems to cease. So is the Appel–Haken work really a proof? Is a proof supposed to be some organic mass that is never quite right, that is constantly being fixed? Not according to the paradigm set down by Euclid 2300 years ago!

There is hardly anything more reassuring than another, independent proof. Paul Seymour and his group at Princeton University found another way to attack the problem (see [RSST] and [SEY]). And in fact today Gonthier (2004) has used a computer-driven "mathematical assistant" to check the 1996 proof.[70] The new proof by Seymour et al also uses a good deal of computer power, but computers are now much faster and more powerful than in the days of Appel and Haken; and the Seymour algorithm seems to be considerably more stable than the old one from the early 1970s. The recent paper [CAH] offers an approach to the four-color theorem that does not require computer assistance.

But it is still the case that nobody can check the Seymour proof, in the traditional sense of "check by a human." The computer is still performing many millions of calculations, and it is not humanly possible to do so much checking by hand—nor would anyone want to! The fact is, however, that over the course of twenty years, from the time of the original Appel–Haken proof to the advent of the Seymour proof, we, as a community of scholars, have become much more comfortable with computer-assisted proofs. There are still doubts and concerns, but this new methodology has become part of the furniture. There are enough computer-aided proofs around (some of them will be discussed in this book) so that a broad cross-section of the community has come to accept them—or at least to tolerate them.[71]

It is still the case that mathematicians are most familiar with, and most comfortable with, a traditional, self-contained proof that consists of a precise sequence of steps recorded on a piece of paper. We still hope that someday there will be such a proof of the four-color theorem. After all, it is only a traditional, Euclidean-style proof that offers the understanding,

---

[70]This process has in fact become an entire industry. There is now a computer verification by Bergstra of Fermat's last theorem, and there is also a computer verification of the prime number theorem (about the distribution of the primes).

[71]It may be noted, however, that Jonathan Borwein and his team have evolved a view of our subject called *experimental mathematics*. The basis of their program is that one can do exploration and experimentation using the computer, but ultimately one writes down a traditional proof. According to Borwein, this paradigm has met with considerable resistance from the mathematical community.

the insight, and the sense of completion that all scholars seek. For now we live with the computer-aided proof of the four-color theorem.

With the hindsight of thirty years, we can be philosophical about the Appel–Haken proof of the four-color theorem. What is disturbing about it, and about the Hales proof of the Kepler conjecture (discussed elsewhere in this book) is that these proofs lack the sense of *closure* that we ordinarily associate with mathematical proof. Traditionally, we invest several hours—or perhaps several days or weeks—absorbing and internalizing a new mathematical proof. Our goal in the process is to *learn something*.[72] The end result is new understanding, and a definitive feeling that something has been internalized and accomplished. These new computer proofs do not offer that reward. The Grisha Perelman proof of the Poincaré conjecture, and the classification of the finite, simple groups, also lack the sense of closure. In those cases computers do not play a role. Instead the difficulty is with the sociology of the profession and the subject. Those proofs are also discussed in these pages.

The real schism is, as Robert Strichartz [STR] has put it, between the quest for knowledge and the quest for certainty. Mathematics has traditionally prided itself on the unshakable absoluteness of its results. This is the value of our method of proof as established by Euclid. But there are so many new developments that have undercut the foundations of the traditional value system. And there are new societal needs: theoretical computer science and engineering and even modern applied mathematics require particular pieces of information and identifiable techniques. The need for a workable device often far exceeds the need to be *certain* that the technique can stand up to the rigorous rules of logic. The result may be that we shall reevaluate the foundations of our subject. The way that mathematics is practiced in the year 2100 may be quite different from the way that it is practiced today.

Doron Zeilberger is a witty and accomplished mathematician who asks this question. Suppose that somebody said, "Here is a proof that the Goldbach conjecture [discussed elsewhere in this book] that is true with 99% certainty. With $10 billion in funding I can come up with a proof that is 100% certain." What are we then to do?

Nobody—not even Bill Gates—is going to pay $10 billion for a definitive solution of the Goldbach conjecture. It is just not that important. So will we learn to live with a 99% sure solution? Or will we consider the problem still open? These are fascinating questions with no clear answer.

---

[72] And the admittedly selfish motivation is to learn some new techniques that will help the reader with his/her own problems.

7

Computer-Generated Proofs

Automatic theorem proving remains a primitive art, able to generate only the most rudimentary arguments.

—Arthur Jaffe

As wider classes of identities, and perhaps even other kinds of classes of theorems, become routinely provable, we might witness many results for which we would know how to find a proof (or refutation); but we would be unable or unwilling to pay for finding such proofs, since "almost certainty" can be bought so much cheaper. I can envision an abstract of a paper, c. 2100, that reads, "We show in a certain precise sense that the Goldbach conjecture is true with probability larger than 0.99999 and that its complete truth could be determined with a budget of $10 billion."

As absolute truth becomes more and more expensive, we would sooner or later come to grips with the fact that few non-trivial results could be known with old-fashioned certainty. Most likely we will wind up abandoning the task of keeping track of price altogether and complete the metamorphosis to nonrigorous mathematics.

—Doron Zeilberger

7.1 A Brief History of Computing

The first counting *devices* were counting boards. The most primitive of these may be as old as 1200 BCE and consisted of a board or stone tablet with mounds of sand in which impressions or marks could be made. Later counting boards had grooves or metal disks that could be used to mark a position. The oldest extant counting board is the Salamis tablet, dating to 300 BCE. See Figure 7.1.

The *abacus* was the first computing machine. Usually credited to the Chinese, it is first mentioned (as far as we know) in a book of the Eastern Han Dynasty, written by Xu Yue in 190 CE. This is a frame, often made of wood, equipped with cylindrical dowels along which beads may slide. See Figure 7.2. In a modern abacus, the beads in the lower portion of the abacus each represent one unit, while the beads in the upper portion each represent five units.

Counting boards were prevalent in ancient Greece and Rome. The abacus developed slowly over a period of 1000 years. The modern abacus was popularized during the Song Dynasty in China during the period 960–1127. In the succeeding centuries the use of the abacus spread to Japan and Korea. It is still in common use today. In fact in Taiwan and China one sees shopkeepers using an abacus to check the accuracy of the electronic cash register!

The first mechanical calculator was devised by Blaise Pascal (1623–1662), a mathematician who is remembered today for his work in probability theory and for his philosophical writings. Pascal built his machine in 1642, inspired by a design of Hero of Alexandria (c. 10–70 CE). Pascal was interested in adding up the distance that a carriage traveled.

Pascal's design is still in use in water meters and automobile odometers. It is based on a single-tooth gear engaged with a multi-tooth gear (Figure 7.3). The basic idea was that the one-tooth gear was large enough to engaged only the multi-tooth gear after one unit of distance had been traversed by the carriage. None other than Gottfried Wilhelm von Leibniz (1646–1716) augmented Pascal's calculator so that it could multiply. In fact, multiplication for this machine consisted of multiple additions. It is possible that technophobia was instigated by Pascal's invention, for even then mathematicians feared for their careers because of this new machine.

Thomas of Colmar (1785–1870) invented the first genuine mechanical calculator in 1820. It could add, subtract, multiply, and divide. Charles Babbage

Figure 7.1.   The Salamis tablet.

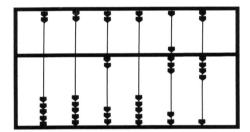

Figure 7.2.   An abacus.

Figure 7.3.   Pascal's gears.

(1791–1871) capitalized on Colmar's idea by adding an insight of his own. Babbage realized as early as 1812 that many long calculations involved repetitive steps. He reasoned that one should be able to build a calculating machine that could handle those redundant steps automatically. He built a prototype of his "difference engine" in 1822. Soon thereafter he obtained a subsidy from the British government to develop his idea. In 1833, however, Babbage got an even better idea.

The better idea was to produce an "analytical engine" which would be a real parallel decimal computer. This machine would operate on "words" of 50 decimal places and was able to store 1000 such numbers. The analytical engine would have several built-in mathematical operations, and these could be executed in any order (as specified by the operator). The instructions for the machine were stored on punch cards (this idea was inspired by the Jacquard loom).

An interesting historical side note is that Lord Byron's daughter Augusta (1815–1852) played a role in Babbage's work. She was a skilled programmer (perhaps the first ever!), and created routines for his machine to calculate the Bernoulli numbers. Lord Byron is remembered as a poet, and he celebrated his daughter with these words:

> My daughter! with thy name this song begun—
> My daughter! with thy name thus much shall end—
> I see thee not,—I hear thee not,—but none
> Can be so wrapt in thee; thou art the friend
> To whom the shadows of far years extend:
> Albeit my brow thus never shouldst behold,
> My voice shall with thy future visions blend,
> And reach into thy heart,—when mine is cold,—
> A token and a tone even from thy father's mould.

In 1869, W.S. Jevons (1835–1882) designed and commissioned the construction of a "logical piano." The purpose of this machine, involving a keyboard and a system of rods, was to perform automated syllogistic reasoning. In fact the machine could do automated deductions of conclusions from premises in simple syllogistic reasoning.

In 1890, Herman Hollerith (1860–1929) created a machine that would read punch cards that he produced. He was working for the U.S. Census Bureau, and he used the machine to tabulate census data.[73] His "tabulating machine" was a terrific innovation, because it increased accuracy and reliability by a dramatic measure. Also the cards were a convenient and reliable way to store the data.[74] In fact Hollerith's tabulator became so successful that he started his own firm to market the device; this company eventually became International Business Machines (or IBM).

Hollerith's machine was limited in its abilities; it could do only tabulations, not direct more complex computations. In 1936, Konrad Zuse (1910–1995) produced a machine (called the Z1) that could be termed the first freely programmable computer. The point here is that this was the first computing machine that was not dedicated to a particular task. Like a modern computer, it could be programmed to do a variety of things. The Z1 was succeeded

---

[73]In fact, after the census of 1880, it took nine years to tally the results. This was before the days of Hollerith, who cut down the processing time by an order of magnitude.

[74]An interesting historical footnote is that after Hollerith started his company, he charged for the cards—*not* for the machines. This is reminiscent of some of the modern printer manufacturers who make all their money selling toner cartridges.

in 1941 by a more advanced machine (called the Z3) that might be termed the first stored-program computer. It was designed to solve complex engineering problems. The machine was controlled by perforated strips of discarded movie film. It should be noted however that, unlike a modern computer, the Z3 was not all-electronic. It was mostly mechanical.

One remarkable feature of Zuse's Z3 is that it used the binary system for numbers. Thus all numbers were encoded using only 0s and 1s (as an instance, the number that we call 26 would be represented as 11010). Zuse may have been inspired by Alan Turing's "Turing machine." In any event, the binary system is used almost universally on computers today.

The year 1942 saw the premiere of the world's first electronic-digital computer by Professor John Atanasoff (1903–1995) and his graduate student Clifford Berry (1918–1963), of Iowa State University. Their computer was called the ABC Computer, and it featured—among its many innovations—a binary system of arithmetic, parallel processing, regenerative memory, and a separation of memory and computing functions. Atanasoff was awarded the National Medal of Science and Technology in 1990 in recognition of his work.

During the years 1939–1944, Howard Aiken (1900–1973) worked in collaboration with engineers at IBM to produce a large, automatic digital computer (based on standard IBM electromechanical parts). The machine, officially known as the "IBM automatic sequence controlled calculator," or ASCC, soon earned the nickname "Harvard Mark I." A remarkable piece of technology, the Harvard Mark I had 750,000 components, was 50 feet long, 8 feet tall, and weighed about 5 tons. It manipulated numbers that were 23 digits long; in fact the machine could add or subtract two such numbers in 0.3 seconds, multiply two such numbers in 4 seconds, and divide them in 10 seconds.

The Harvard Mark I read data from punch cards and received instructions from a paper tape. It consisted of several different calculators that worked on different parts of the problem (much like a parallel-processing machine today). Aiken went on to develop many computing machines, but the Harvard Mark I was perhaps the most significant. It was historically important, and was still in use at Harvard as late as 1959.

As an engineer, Howard Aiken was farseeing and innovative. As a social engineer, perhaps less so. For example, he predicted that six computers would be adequate in the future for all the computing needs in the United States.

Contemporary with the work we have been describing are the influential ideas master-minded by Alan Turing (1912–1954) in Great Britain. In 1936 Turing penned a paper called *On computable numbers*, in which he described a theoretical device now known as a *Turing machine*.

A *Turing machine* is a device for performing effectively computable operations. It consists of a machine through which a bi-infinite paper tape is fed. The tape is divided into an infinite sequence of congruent boxes (Figure 7.4). Each box has either a numeral 0 or a numeral 1 in it. The Turing machine has finitely many "states" $S_1, S_2, \ldots, S_n$. In any given state of the Turing machine, one of the boxes is being scanned.

After scanning the designated box, the Turing machine does one of three things:

(1)   It either erases the numeral 1 that appears in the scanned box and replaces it with a 0, or it erases the numeral 0 that appears in the scanned box and replaces it with a 1, or it leaves the box unchanged.

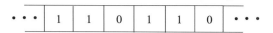

Figure 7.4. A Turing machine.

(2) It moves the tape one box (or one unit) to the left or to the right.

(3) It goes from its current state S_j into a new state S_k.

Turing machines have become a model for anything that can be effectively computed. The theory of recursive functions and the famous "Church's thesis" from formal logic both have interesting and compelling interpretations in terms of Turing machines.

It is safe to say that by 1945 Alan Turing had all the key ideas for what would become the modern stored program computer. He knew that there could be a single machine that could perform all possible computing tasks, and that the key to this functionality was the program stored in memory. Turing was virtually unique in that he understood the mathematical theory behind computing but he also had hands-on experience with large-scale electronics.

The next American breakthrough in the computer race came in 1946 from John W. Mauchly (1907–1980) and J. Presper Eckert (1919–1995) at the University of Pennsylvania. Known as the ENIAC, their computer used a record 18,000 vacuum tubes. Vacuum tubes (which have now been superseded by solid-state devices) generate a good deal of heat, and the cooling machinery for the ENIAC occupied 1800 square feet of floor space. The computer itself weighed 30 tons.

The ENIAC had punch card input and was able to execute complex instructions. The machine had to be reconfigured for each new operation. Nevertheless, it is considered by experts to have been the first successful high-speed electronic digital computer. It was used productively for scientific calculation from 1946 to 1955. Eckert and Mauchly went on to form a computer company in Philadelphia that produced, among other innovations, the UNIVAC computer. This was the first commercially available computer in the United States. Curiously, a patent infringement case was filed in 1973 (Sperry Rand vs. Honeywell) that voided the ENIAC patent as a derivative of John Atanasoff's invention.

The year 1947 saw another computing innovation from Tom Kilburn (1921–2001) and Frederic Williams (1911–1977) of the University of Manchester in England. They succeeded in developing a technology for storing 2048 bits of information (a *bit* is a unit of computer information) on a CRT (cathode-ray tube). They built a computer around this device which became known as the Williams Tube. More precisely, the terminal mirrored what was happening within the Williams Tube. A metal detector plate, placed close to the surface of the tube, detected changes in electrical discharges. Since the metal plate would obscure a clear view of the tube, the technicians could monitor the tube with a video screen. Each dot on the screen represented a dot on the tube's surface; the dots on the tube's surface worked as capacitors that were either charged and bright or uncharged and dark. The information translated into binary code that became a way to program the computer.

By 1948 the Tube had evolved into a new machine called the Baby. Its innovative feature was that it could read and reset at high speed random bits of information, while preserving a bit's value indefinitely between resettings. This was the first computer ever to be able to hold any (small) program in electronic storage and process it at electronic speeds. Kilburn and Williams continued to develop their ideas, and this work led to the Manchester Mark 1 in 1949. In fact, that computer contained the first random access memory (analogous to the sort of memory used in computers today). The point is that data was not stored in sequence—as on a tape. Instead it was stored in such a way that it could be electronically "grabbed" at high speed.

John von Neumann (1903–1957) had a chance encounter in 1944 with Herman Goldstine (1913–2004) at a train station in Aberdeen, Maryland. There von Neumann learned of the EDVAC (Electronic Discrete Variable Automatic Computer) project at the University of Pennsylvania. He became involved in the project, and in 1945 produced a paper that would change the course of the development of computer science. Until then, computers (such as the ENIAC) were such that they had to be physically reconfigured for each new task. Von Neumann's vision was that the instructions to the computer could be stored *inside the machine*, and he was the first to conceive of a *stored program computer* as we think of it today. Ultimately von Neumann and Goldstine moved back to the Institute for Advanced Study (IAS) in Princeton, New Jersey, where they developed the IAS computer which implemented von Neumann's ideas. Many consider von Neumann (a mathematician by pedigree) to be the father of the modern computer.

In 1956, A. Newell (1927–1992) and H. A. Simon (1916–2001) created the Logic Theory Machine which could prove theorems in the propositional calculus (the part of formal logic that deals with the relationships between propositions). The Logic Theory Machine could in practice find only very short proofs. This machine was quickly succeeded by the Geometry Machine of H. Gelernter. By restricting attention to a particular part of mathematics (geometry), Gelernter was able to achieve greater efficiency.

The year 1956 also saw the start of a project to automate the proofs of all the results in Whitehead and Russell's *Principia Mathematica* using the Logic Theory Machine of Newell and Simon. This project was completed in 1959.

By 1960, there were three prominent theorem-proving technologies that had been created by P. Gilmore, H. Wang (1921–1995), and D. Prawitz (1936–   ). These systems could handle only very elementary statements and proofs.

In the late 1960s a group at the Applied Logic Corporation in Princeton, New Jersey, developed a system called SAM (semi-automated mathematics). Their innovation was that the system functioned with *human intervention and guidance*.

In 1967 the proof-checking system Automath was developed at the Technische Hogeschool, Eindhoven by N.G. de Bruijn and his colleagues. It had its own special language for recording and checking mathematics.

In 1972 R. A. Overbeek created the theorem-proving system called Aura (Automated Reasoning Assistant). It has since been replaced, and surpassed, by Otter. Other modern and powerful systems for doing mathematical proofs on computer are HOL Light (where HOL stands for "Higher Order Logic"), Mizar, ProofPower, Isabelle, and Coq. The articles [HAL3] and [WIE] provide more information about these tools.

Some notable triumphs for computer theorem-proving are these:

- **The First Gödel Incompleteness Theorem** This is the statement that, in any sufficiently sophisticated logical system (one that contains arithmetic), there are true statements that cannot be proved. See our Section 7.10 for further details. Shankar used the computer utility `ngthm` to provide a "formal proof" (meaning a computer-generated proof from the most basic axioms, with a tedious step-by-step logical procedure) of this result in 1986. O'Connor used `Coq` in 2003 to provide an even more streamlined proof, and Harrison used `HOL Light` in 2005 to generate another proof.

- **The Jordan Curve Theorem** This is the result that a closed, non-self-intersecting curve in the plane divides the plane into two regions—a bounded one and an unbounded one. Intuitively obvious, this theorem is incredibly difficult to prove in the traditional sense of the word. In 2005, Thomas Hales used `Mizar` to produce a formal proof of this fact.

- **Prime Number Theorem** This is the celebrated result of de la Vallée Poussin and Hadamard about the distribution of prime numbers. It is a hard proof, and uses complex analysis. In 2008 John Harrison used `HOL Light` to produce a formal proof.

- **The Four-Color Theorem** We discuss the four-color theorem in some detail in Chapter 6. The original proof of this result by Appel and Haken used the computer to do millions of calculations. Now there is a formal proof, produced in 2004 using `Coq`, by Georges Gonthier. See the article [GON] for more on this event.

It must be understood that the computer creation of a formal proof can be quite a tedious and complex procedure. For instance, A. Matthias has calculated that to expand the definition of the number 1 fully in terms of logical primitives requires over 4 trillion symbols. So imagine what it would take to fully formalize a complicated mathematical proof such as the four we have just described!

An interesting byroad in the modern history of computing is the supercomputer. Seymour Cray (1925–1996) was arguably the godfather of supercomputing. Super-computers, in Cray's vision, are founded on the notion of "parallel processing." The idea is that every computer contains a central processing unit (CPU). This is the chip in which all the calculations—the manipulations of 0s and 1s—take place. A supercomputer will have several CPUs—sometimes more than 100 of them[75]—and the different computing tasks will be parceled out among them. Also each CPU will have its own memory bank—often comprising several gigabytes.

Therefore, many different calculations can be taking place at the same time. The consequence is that the end result is reached more quickly. In the early days of supercomputing, the user had to learn a special programming language that had routines for doing this division of labor among the different processors. In today's supercomputers, the parceling is done automatically, just by parsing the standard code that the user inputs. There are versions of `Fortran` and `C` and `C++` that can be used on supercomputers. Other widely used programming environments on supercomputers are Message Passing Interface (`MPI`), `OpenMP`, Co-Array Fortran, and Universal Parallel `C`.

---

[75]In fact, the current record is that there is a single supercomputer with 212,992 CPUs.

For many years, the fastest parallel-processing machines came out of Seymour Cray's company Cray Research, outside of Minneapolis. And many of the hottest new high tech companies were spin-offs from Cray. For quite a time, the speed of a Cray was the benchmark for the computing world. As a simple example, when the first Pentium chips came out people crowed that "now we have a Cray I on a chip." That chip ran at 100 megaflop/s (or 100 Mflop/s).[76]

Seymour Cray was a special and enigmatic man. When he had a deep and difficult problem to think about, he would go underground. Literally. For many years, Cray worked on digging—with a hand shovel—a $4' \times 8'$ tunnel that was to connect his house in Chippewa Falls with a nearby lake. Cray said he always did his best thinking while he was shoveling away at his tunnel—"While I'm digging in the tunnel, the elves will often come to me with solutions to my problem." He died in an automobile accident in Colorado before the tunnel was completed.

Today, with the advent of the *personal computer*, computing has become a major feature of life for all of us. Steve Jobs (1955–) and Steve Wozniak (1950–) invented the mass-market personal computer in 1977.[77] IBM revolutionized the personal computer industry in 1981 with the introduction of the first PC. This machine used an operating system that was first developed as 86-DOS by Tim Paterson[78] of the Seattle Computer Company. Bill Gates of Microsoft purchased 86-DOS for $50,000 and turned it into MS-DOS, which was the operating system for the IBM PC from 1981 until 1995.

The year 1994 saw the development of the modern, sophisticated theorem-proving software Otter by W. W. McCune. Otter has been used not just to "rediscover" proofs that are already known, but also to discover genuinely new mathematical theorems and their proofs. Otter has now been superseded by Prover9.

---

[76] A *flop* is, in computer jargon, a "floating-point operation." The acronym *flop/s* stands for "floating-point operations per second." A flop is one elementary arithmetic calculation, like addition. One *Mflop* is one million floating point operations. So a 100-megaflop machine can perform 100 million elementary arithmetic operations per second. That is the speed of an ordinary desktop computer from 10 years ago. The desktop computers today are five to ten times as fast. The fastest machines in the world today (both supercomputers with parallel processing) are the MDGrape by RIKEN, which calculates at 1 petaflop/s (that is, 10^{15} flop/s, or one quadrillion flop/s, or 1 Pflop/s) and the IBM Blue Gene/L, which calculates at 360 teraflop/s (that is, $360 \cdot 10^{12}$ flop/s, or 360 trillion flop/s, or 360 Tflop/s.

[77] It should be noted, however, that the Commodore PET computer appeared at about the same time as the Apple II. Also a version of the personal computer was produced earlier at the Xerox PARC research facility in Palo Alto. The Xerox PARC personal computer was never a commercial product. Xerox PARC also invented the mouse, the laser printer, and the graphic user interface that has today developed into the operating system Windows.

The operating system called Microsoft Windows is compiled from 100 million lines of computer code. This is one of the most remarkable engineering feats in history. When the merits of President Ronald Reagan's *Star Wars* program were being debated, it was pointed out that the control system for the project would involve tens of thousands of lines of computer code. At that time such complexity was considered to be infeasible. We have clearly evolved far beyond that state of sophistication. The verification of the reliability of something as complex as Windows is tantamount to proving a mathematical theorem. Indeed, T. Ball at Microsoft has developed theorem-proving and model-checking software that is used to verify portions of the Windows operating system. Just to check 30 functions of the Parallel Port device driver, the proving software was invoked 487,716 times.

[78] Tim Paterson now works for Microsoft.

7.2 The Difference between Mathematics and Computer Science

When the average person learns that someone is a mathematician, he or she often supposes that that person works on computers all day. This conclusion is both true and false.

Computers are a pervasive aspect of all parts of modern life. As we learned in the last section, the father of modern computer design was John von Neumann, a mathematician. He worked with Herman Goldstine, also a mathematician. Today, nearly every mathematician uses a computer to do email, to typeset papers and books, and to post material on the World Wide Web. A significant number (but well less than half) of mathematicians use the computer to conduct *experiments*. They calculate numerical solutions of differential equations, calculate propagation of data for dynamical systems and differential equations, perform operations research, engage in the examination of questions from control theory, calculate the irreducible unitary representations of Lie groups, calculate L-functions in analytic number theory, and many other activities as well. But the vast majority of (academic) mathematicians still, in the end, pick up a pen and write down a *proof*. And that is what they publish.

The design of the modern computer is based on mathematical ideas—the Turing machine, coding theory, queuing theory, binary numbers and operations, programming languages, and so forth. Operating systems, high-level computing languages (like `Fortran`, `C++`, `Java`, etc.), central processing unit (CPU) design, memory chip design, bus design, memory management, and many other components of the computer world are mathematics-driven. The computer world is an effective and important implementation of the mathematical *theory* that we have been developing for 2500 years. But the computer *is not mathematics*. It is a device for manipulating data.

Still, exciting new ideas have come about that have altered the way that mathematics is practiced. We have seen that the earliest computers could do little more than arithmetic. Slowly, over time, the idea developed that the computer could carry out *routines*. Ultimately, because of work of von Neumann, the idea of the stored-program computer was developed. In the 1960s, a group at MIT developed the idea that a computer could perform high-level algebra and geometry and calculus computations. Their software product was called `Macsyma`. It could run only on a very powerful computer, and its programming language was complex and difficult.

Thanks to Stephen Wolfram (1959–   ),[79] the Maple group at the University of Waterloo,[80] the MathWorks group in Natick, Massachusetts,[81], and many others, we have *computer algebra systems*. A computer algebra system is a high-level computer language that can do calculus, solve differential equations, perform elaborate algebraic manipulations, graph very complicated functions, and perform a vast array of sophisticated mathematical operations. And these software products will run on a personal computer! A great many mathematicians and engineers and other mathematical scientists conduct high-level research

---

[79] His famous product is `Mathematica`.

[80] Their famous product is `Maple`.

[81] Their famous product is `MATLAB`.

(B_3) $x \cup (x \cap y) = x$ $(\widetilde{B}_3)$ $x \cap (x \cup y) = x$

(B_4) $x \cap (y \cup z) = (x \cap y) \cup (x \cap z)$ $(\widetilde{B}_4)$ $x \cup (y \cap z) = (x \cup y) \cap (x \cup z)$

(B_5) $x \cup \overline{x} = 1$ $(\widetilde{B}_5)$ $x \cap \overline{x} = 0$

In the 1930s, Herbert Robbins conjectured that these ten axioms were in fact implied by just three rather simple axioms:

(R_1) $x \cup (y \cup z) = (x \cup y) \cup z$ (associativity)
(R_2) $x \cup y = y \cup x$ (commutativity)
(R_3) $\overline{\overline{x \cup y} \cup \overline{x \cup \overline{y}}} = x$ (Robbins equation)

It is not difficult to show that every Boolean algebra, according to the original ten axioms (B_1)–(B_5) and $(\widetilde{B}_1)$–$(\widetilde{B}_5)$, is also a Robbins algebra (i.e., satisfies the three new axioms (R_1)–(R_3)). But it was an open question for sixty years whether every Robbins algebra is a Boolean algebra. The question was finally answered in the affirmative by William McCune of the Argonne National Laboratory, using software (developed at Argonne) called EQP (for "Equational Theorem Prover"). The computer was simply able to fit the Robbins axioms together in millions of ways, following the strict rules of logic, and find one that yields the ten axioms of Boolean algebra. It is important to note that after the computer found a proof of the Robbins conjecture, Allen L. Mann [MAN1] produced a proof on paper that can be read by a human being.

Computers have been used effectively to find new theorems in projective geometry and other classical parts of mathematics. Even some new theorems in Euclidean geometry have been found (see [CHO]). Results in algebra have been obtained by Stickel [STI]. New theorems have also been found in set theory, lattice theory, and ring theory. One could argue that the reason these results were never found by a human being is that no human being would have been interested in them. Only time can judge that question. But the positive resolution of the Robbins conjecture is of great interest for theoretical computer science and logic.

One of the real pioneers in the search for computer proofs is Larry Wos, of Argonne National Laboratory. He is a master at finding new proofs of unproved theorems, and also of finding much shorter proofs of known theorems. He does all this on the computer, often using the software Otter by W. McCune. One of Wos's recent coups is the computer-generated solution of the SCB problem in equivalential calculus. Wos's work is well archived in various books, including [WOS1] and [WOS2].

7.5 How the Computer Generates the Proof of a New Result

There are now dedicated pieces of software that can search an axiom system for new results. For example, the program Otter is an automated reasoning program. You can tell it the problem you want it to think about and what type of logic to employ and then off it goes.

It has been used successfully to analyze a number of different situations in many different fields of mathematics.

As indicated in the last section, the Robbins conjecture in Boolean algebra is an outstanding example of an important new fact that was discovered with a computer search. It should be understood that the axioms of Boolean algebra are crisp, neat, and fit together like building blocks. This is a system of analytical thinking that lends itself well to this new technology of computer-aided theorem-proving. Subject areas such as real analysis, and partial differential equations, and low-dimensional topology—which use a great deal of synthetic argumentation and proof-by-picture and ad hoc argument—do not (at least with our current level of knowledge in computer proofs) lend themselves well to computer searches for proofs. There are those, such as Doron Zeilberger [ZEI], who predict that in 100 years *all* proofs will be computer-generated. Given the current state of the art, and the evidence that we have at hand, this seems like wishful thinking at best.

One interesting train of thought that has developed is that many computer proofs are of necessity too long and complex to be of any use to human beings. In 1972, Albert Meyer of MIT showed that computer proofs of some arbitrarily chosen statements in a very simple logical system (in fact a system in which the only operation is to add 1 to an integer) are infeasibly long. In later work it was shown that in this same very simple logical system, there are statements with 617 or fewer symbols that require 10^{123} components. Note that 10^{123} is the number of proton-sized objects that would densely fill the known universe. So the lesson is that while computer proofs are interesting, and continue to reveal new truths, they have their limitations. It seems unlikely that, someday, computers will be generating all the proofs that we need.

It must be recorded that the fact that a computer can perform a high-level task like searching for a new truth and then searching for its proof is really quite impressive. Consider that computers were first constructed—about sixty years ago—to do calculations of the weather and of artillery data. All the calculations were numerical, and they were quite elementary. The point of computers in those days was *not* that the computer could do anything that a person could not do. Rather, it was that the computer was faster and more accurate.

There is an amusing story that may tend to *gainsay* the point made in the last paragraph. As noted elsewhere in this book, von Neumann conceived and developed the first stored-program computer at the Institute for Advanced Study in Princeton. He had a whole team of people, including Herman Goldstine, working with him on that project.

John von Neumann was an amazing mathematician and an amazing mental calculator. He also had a photographic memory: he could effortlessly recite long passages from novels that he had read twenty years earlier. In addition to developing (along with Herman Goldstine) one of the first stored-program computers, von Neumann had an active life as a mathematician and a consultant. He was constantly coming and going, all over the country, to government agencies and companies, disseminating the benefit of his erudition. It was said that his income from these activities (on top of his princely Institute salary) was quite substantial. And he was already rather wealthy with family money.

During one of von Neumann's consulting trips, Herman Goldstine and the others working on the new computer got it up and running for a test. They fed it a large amount of data from meteorological observations, ran it all night, and came up with very interesting conclusions in the morning. Later that day, von Neumann returned from his trip. Wanting to pull a prank on von Neumann, they decided not to tell him that they had the computer up and running, but instead to present their results as though they had obtained them by hand. At tea, they told von Neumann that they had been working on such and such a problem, with such and such data, and in the first case had come up with . . . . "No, no," said von Neumann. He put his hand to his forehead, threw his head back, and in a few moments gave them the answer. It was the same answer that the machine had generated. Then they said, "Well, in the second case we got . . . . "No, no," said von Neumann. "Let me think." He threw his head back—it took longer this time—but after several moments he came up with the answer. Finally his collaborators said, "Now in the third case . . . ." Again, von Neumann insisted on doing the calculation himself. He threw his head back and thought and thought and thought. After several minutes he was still thinking and they blurted out the answer. John von Neumann came out of his trance and said, "Yes, that's it. How did you get there before I did?"

Another story of von Neumann's calculating prowess goes like this. He was at a party, and someone asked him the following chestnut:

> Two trains are fifty miles apart. They travel toward each other, on the same track, at a rate of 25 miles per hour. A fly begins on the nose of one train, travels at a rate of 40 miles per hour back to the first train. The fly travels back and forth between the trains until it is crushed between the trains. How far does the fly travel in total?

Johnny von Neumann instantly blurted out the correct answer of 40 miles. His friends said, "Oh, you saw the trick." (The trick is to notice that it takes the trains one hour to meet. Hence the fly travels for one hour, or 40 miles.) But von Neumann said, "No, I just added up the infinite series."

It is fun—if extremely tedious—to note what that series actually is. First observe that if the situation is as in Figure 7.5, with the fly beginning on the left-hand train (located at the origin), then we can calculate when the fly first meets the right-hand train with the equation

$$40t = 50 - 25t \, .$$

Thus the first flight of the fly, from the left-hand train to the right-hand train, has duration $t = 50/65$. The second flight of the fly is found by solving

$$\frac{50 \cdot 40}{65} - 40t = \frac{50}{65} \cdot 25 + 25t,$$

hence that second flight has duration $t = (50 \cdot 15)/65^2$. Continuing in this manner, we end up with the infinite series of times

$$T = \frac{50}{65} + \frac{50}{65^2} \cdot 15 + \frac{50}{65^3} \cdot 15^2 + \frac{50}{65^4} \cdot 15^3 + \cdots$$

$$= \frac{50}{65} \left(1 + \frac{15}{65} + \left(\frac{15}{65} \right)^2 + \left(\frac{15}{65} \right)^3 + \cdots \right) .$$

Figure 7.5.   The fly traveling between two trains.

Now one may use standard techniques to sum the infinite series in the large parentheses and find that the answer is 65/50. Thus the total elapsed time before the fly is crushed is 1 hour. In conclusion, the fly travels a total of 40 miles.

Today even a personal computer can run at a speed of up to one gigaflop/s (or Gflop/s). So the computer is performing 1 billion basic arithmetic operations per second. Clearly no man—not even John von Neumann—can match that speed.

In the next chapter we begin to explore new ways that computers are being used in communication, in teaching, and in proof. There is no denying that computers have changed many aspects of our lives, and they continue to do so.

8

The Computer as an Aid to Teaching and a Substitute for Proof

Light; or, failing that, lightning: the world can take its choice.

—Thomas Carlyle

It takes a long time to understand nothing.

—Edward Dahlberg

Experimental mathematics is that branch of mathematics that concerns itself ultimately with codification and transmission of insights within the mathematical community through the use of experimental exploration of conjectures and more informal beliefs and a careful analysis of the data acquired in this pursuit.

—J. Borwein, P. Borwein, R. Girgensohn, and S. Parnes

When a computer program applies logical reasoning so effectively that the program yields proofs that are published in mathematics and in logic journals, an important landmark has been reached. That landmark has been reached by various automated reasoning programs. Their use has led to answers to open questions from fields that include group theory, combinatory logic, finite semigroup theory, Robbins algebra, propositional calculus, and equivalential calculus.

—Keith Devlin

8.1 Geometer's Sketchpad

Geometer's Sketchpad is a (software) learning tool, *marketed by Key Curriculum Press*, designed for teaching Euclidean geometry to high school students. There has been a great trend in the past twenty-five years to reverse-engineer high school geometry so that proofs are deemphasized and empiricism and speculation more highly developed. Geometer's Sketchpad fits in very nicely with this program.

Geometer's Sketchpad enables the user to draw squares and triangles and circles and other artifacts of classical geometry and to fit them together, dilate them, compare them, measure congruences, and so forth. It is a great way to experiment with geometrical ideas. And, perhaps most important, it is an effective device for generating student interest in learning mathematics. Students today are loathe to read the dry text of a traditional mathematical proof. They are much more enthusiastic about jumping in and doing experiments with `Geometer's Sketchpad`. Thus this new software can, in the right-hands, be a dynamic and effective teaching tool. And it has been a great success in the marketplace. Entire countries—most recently Thailand—have purchased site licenses for `Geometer's Sketchpad`.

8.2 Computer Algebra Systems

An exciting new development in the mathematical infrastructure of the past thirty-five years has been *computer algebra systems (CAS)* (also known as symbol manipulation systems (SMS)). Whereas, in the early days, computers were devices for crunching numbers—the users input a lot of numerical data and the machine spat out a lot of numerical data—the premise of a CAS is that the machine can *actually do mathematics*. This might consist of solving a system of algebraic equations, finding roots of polynomials or functions, graphing complicated functions of one or several variables, solving differential equations, calculating derivatives and integrals, doing statistical calculations, modeling, and carrying out many other mathematical activities. There are a good many products, both commercial and freeware, available today for various CAS activities. Some of these are "general purpose" software that can perform a vast range of mathematical functions. Others are dedicated to a particular collection of activities, such as algebra or statistics or geometry.

The software `Macsyma`, developed from 1967 to 1982 at MIT by William A. Martin, Carl Engelman, and Joel Moses, was revolutionary because it was the first all-purpose *symbolz manipulation* package. `Macsyma` offered the capability to perform *algebraic operations*. `Macsyma` could solve systems of equations, calculate integrals, invert matrices, solve differential equations, compute eigenvalues, and do a number of other calculations that involve *symbols* rather than numbers.

The somewhat daunting feature of `Macsyma` was that it ran only on a fairly powerful computer, and its programming language was in the nature of an artificial intelligence language. It was difficult to program in `Macsyma`. But it was all we had for a number of years. It would be quite difficult, for example, to calculate the eigenvalues of a 20×20 matrix by hand; `Macsyma` can do it in a trice. It would be nearly prohibitive to solve a system of ten ordinary differential equations in ten unknowns by hand; for `Macsyma` the matter is straightforward. Even though there are a number of software products today that have superseded `Macsyma`, there are legacy routines—developed, for instance, to study questions of general relativity—that were written in `Macsyma`. So people continue to use the product.

In the early 1980s, the nature of symbolic manipulation changed dramatically. MacArthur Fellowship recipient Stephen Wolfram developed a new package called `Mathematica`. This new tool had many advantages over `Macsyma`:

- Mathematica will run on a microcomputer,[83] for example on a PC or a Macintosh.
- Mathematica has a very intuitive and transparent syntax. Anyone with some mathematical training will find programming in Mathematica to be straightforward.
- Mathematica is very fast. It can do calculations to any number of decimal places of accuracy very quickly.
- Mathematica is a wizard at graphing functions of one or two variables. The user just types in the function—no matter how complicated—and the graph is produced in seconds. Graphs can be viewed from any angle, and they can be rotated in space. (Macsyma, at least in its early incarnations, did not do graphing at all. More recent releases have additional functionality.)[84]
- Mathematica is a whiz at displaying data.

Stephen Wolfram was quite aggressive about marketing his new product, but the fact is that it virtually sold itself (see also Section 7.2). If ever there was a new tool that allowed us to see things that we could not see before, and calculate things that we had not dreamed of calculating before, then this was it. The product Mathematica was and continues to be rather expensive, and the licensing policies of Wolfram Research (the company that produces Mathematica) rather restrictive. But Mathematica has sold like hotcakes. Figure 8.1 exhibits a Mathematica graphic.

Some competing products have come about that give Mathematica a good run for its money. These include Waterloo Maple and MATLAB by MathWorks. Each offers something a little different from Mathematica. For example, many people prefer the syntax of Maple to that of Mathematica. Also Maple is deemed to be more reliable, and offers many functions that Mathematica does not have. Engineers prefer MATLAB, because it is more of a numerical engine than the other two products. And MATLAB has a Maple kernel! So it shares many of the good features of Maple.

It happens that MATLAB is particularly adept at handling complex numbers; Mathematica and Maple are rather clumsy in that context. It has become fashionable to refer to all these software packages as "computer algebra systems." But they are really much more than that.

There are some remarkable products that are built *atop* some of these computer algebra systems. For example, Scientific WorkPlace by Makichan Software is a very sophisticated word processing system. It is particularly designed for producing mathematics documents. Suppose that you are writing a textbook using Scientific WorkPlace. And you display an example. It turns out that Scientific WorkPlace has a Maple kernel, and it will work the example for you! This is a terrific accuracy-checking device. And a great convenience for authors and other working mathematicians.

Here, in chart form, we give very brief descriptions of several CAS software products. The chart gives a sense of the span of time over which these software products have been available, and also the great variety of products that there are. We do not attempt to say anything about price, availability, or licensing. Also our list is not complete; but it gives a

---

[83] More recent releases of Macsyma will also run on a microcomputer.

[84] Wolfram is a millionaire many times over from the sales of Mathematica. He claims that his biggest customers are the business schools, because they appreciate the graphing capabilities of Mathematica.

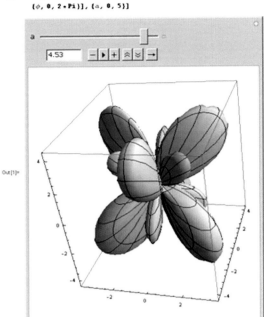

Figure 8.1. A `Mathematica` graphic.

representative sampling of the utilities that are out in the world today. A terrific resource for comparison of these products is

{`http://en.wikipedia.org/wiki/Comparison_of_computer_algebra_systems`}

A Chronological List of Computer Algebra Systems

Event	Purpose	Release Date
Schoonship	particle physics	1963
Macsyma	general purpose	1968
Reduce	general purpose	1968
bergman	algebra	1972
SAS	statistics	1976
MuMATH	general purpose	1980
MATLAB	general purpose	1984
Maple	general purpose	1985
MathCad	general purpose	1985
Derive	general purpose	1988

Event	Purpose	Release Date
PART/GP	number theory, arbitrary precision	1985
MicroMath	curve fitting, data analysis	1985
GAP	computational discrete algebra	1986
Mathematica	general purpose	1988
MuPAD	general purpose	1992
MAGMA	algebra, number theory	1993
GeomView	geometry	1996
Macaulay	algebraic geometry, commutative algebra	2002
SAGE	general purpose	2005

Today many mathematicians keep a CAS always at their side, available and booted up. They routinely perform calculations, carry out what-if experiments, and graph functions in order to help them understand their work. With a CAS one can obtain an extremely accurate graph of a function of two variables in just a few seconds—such a graph may take hours if done by hand, and would have nowhere near the accuracy.

Mathematicians can create instructional Mathematica *Notebooks* to use in class: students can benefit from the computing power of *Mathematica* without learning the technical computer language. The Notebook interfaces with the user using just ordinary English together with point-and-click. Maple has a similar utility.

It would be misleading to suggest that the activities described here are any substitute for traditional mathematical proof. But they are a terrific aid to *searching for a proof*. New insights, new ideas, and new perspectives are often gained from examining CAS calculations and graphics. Many new Ph.D. theses are written based on CAS work. The calculations in the papers [CHE1], [CHE2], although ultimately worked out by hand, were done with the aid of CAS calculations. In hindsight we can say that the CAS calculations showed us what was going on, and helped us to establish a pattern. Then we could forge ahead and do the calculations and prove the theorems in the more traditional fashion.

It is enlightening and thought-provoking to think about what computer calculations can tell us. Consider the identity

$$\frac{24}{7\sqrt{7}} \int_{\pi/3}^{\pi/2} \log \left| \frac{\tan t + \sqrt{7}}{\tan t - \sqrt{7}} \right| dt$$

$$= \sum_{n=0}^{\infty} \left[\frac{1}{(7n+1)^2} + \frac{1}{(7n+2)^2} - \frac{1}{(7n+3)^2} + \frac{1}{(7n+4)^2} \right.$$

$$\left. - \frac{1}{(7n+5)^2} - \frac{1}{(7n+6)^2} \right].$$

There is no known proof of this equation. But it has been verified to 20,000 digits with 45 minutes of computing time on a parallel-processing machine at Virginia Polytechnic

designers would discuss the design of a new auto body, and produce various sketches and color renditions—these were paper hard copies. Then a clay model of the auto body would be produced. At that stage the engineers examined the new design and critiqued it: Where would the engine go? Do the passengers have sufficient room? Is it safe? Is it aerodynamic? And so forth. Their comments and criticisms would be conveyed to the artists and designers, and then new sketches and a new clay model would be produced. Then the engineers would have another go at it. This loop would be repeated a great many times until a workable model was created. The entire process took thousands of person hours and many months. In the early 1980s, Volvo decided that it wanted to modernize and automate the process. The company gave the problem to its in-house technical staff for development. Unfortunately they found it too difficult; they could make no progress with the task.

Björn Dahlberg—also a Swede—was a brilliant theoretical mathematician, *holder of the prestigious Salem Prize* in harmonic analysis. And he took a real shine to the problem of designing automobile bodies. He applied sophisticated techniques from differential equations, differential geometry, linear programming, convex surface theory, harmonic analysis, real variable theory, and other parts of mathematics to help develop the elaborate software package SLIP that Volvo now uses every day to design new autos. The key insight of SLIP is that an auto body is the union of convex surfaces. The user inputs data points through which the surface will pass (or at least "approximately pass"), specifies criteria for air resistance, refraction of light, and other desiderata, and then SLIP does an elaborate calculation and produces the surface—both graphically and analytically. The entire subject of surface design was revolutionized by SLIP.

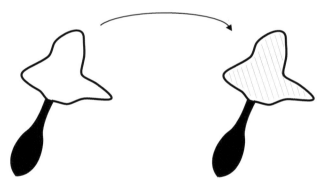

Figure 8.3. A soap film.

Similarly, computer visualization can be an important tool in proofs. Hoffman, Hoffman, and Meeks [HHM] and their geometry group GANG (Geometry Analysis Numerics Graphics) studied embedded minimal surfaces in 3-dimensional space. Here a minimal surface is (locally) a surface of least area. A classical means of creating a minimal surface is to bend a wire to a given shape, and then dip the wire in soap solution. The resulting "soap film" (predicted and described, at least in principle, by the minimal surface differential equations) is a minimal surface. See Figure 8.3. If the wire is in the shape of a circle, then

Differential Geometry

Classical (Euclidean) geometry concerns the relationships among circles, triangles, squares, and other planar figures. But our world is three-dimensional, and is made of up of surfaces and solids that come in a bewildering variety of shapes. *Differential geometry*, a subject pioneered by Carl Friedrich Gauss and Bernhard Riemann, provides analytic tools for understanding the shapes of such objects. The differential geometry of Riemann was crucial to Albert Einstein's formulation of his general relativity theory. Today differential geometry is used in all parts of cosmology and mathematical physics.

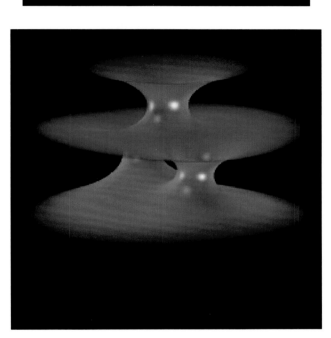

Figure 8.4. A minimal surface.

the resulting minimal surface is a disk. If the wire is in a more exotic shape, then the minimal surface will be a more complicated 2-dimensional configuration.

Hoffman, Hoffman, and Meeks would first generate pictorial examples using methods of numerical analysis and then, after analyzing these examples and determining various patterns and relationships, would write down a traditional theorem and prove their result in a traditional manner. The point here is that the computer enabled these scientists to *see things* that they could not otherwise. It enabled them to envision certain important examples that pointed to new directions in the field. Then, equipped with new understanding and insight, they were able to write down a theorem and prove it. Figure 8.4 exhibits a minimal surface that these three mathematicians have helped us to discover and understand.

The notion of using the computer to develop visualization techniques is a relatively new one. Computer graphic calculations are "computationally expensive"—meaning that they require considerable processing power, considerable memory, and considerable hard disk space, not to mention a good chunk of computing time. Even twenty-five years ago, one was at the mercy of the computer system at the university. One had to compete with computer scientists and engineers and other high-profile users for a time slice on the mainframe (or perhaps the minicomputer in later years) so that the necessary calculations could be performed. Some mathematicians developed the habit of going into work very late at night, often after midnight in order to be able to do their work. Now we have desktop microcomputers that perform at the speed of some of the early supercomputers. My present notebook computer is *much* faster than the Cray I, one of the benchmark machines produced by Seymour Cray in the late 1960s and early 1970s. It operates at considerably greater than 100 Mflop/s. It is only recently that such powerful tools have become readily available and relatively inexpensive.

8.5 Mathematical Communication

In the ancient world the great mathematicians had schools. Euclid, for example, ran a large and influential school in Alexandria. Archimedes had a school. In this way the powerful scholar could disseminate his ideas, and also train young people to help him in his work.

But there was not a great deal of communication among mathematicians. Travel was difficult, and there was no postal system. In many ways mathematicians worked in isolation. An extreme illustration of the lack of communication is that the ancient Chinese anticipated by several centuries the Pythagorean theorem and many other results that we have commonly attributed to Western mathematicians. Yet this fact was not learned until modern times. In fact, the Chinese were rather sophisticated mathematicians in a number of ways that are still not known in the west. In the third century CE, Liu Hui managed to inscribe a 3072-sided polygon in a circle, and thereby to calculate the number π to five decimal places. In the fifth century CE, the father–son team of Tsu Ch'ung-Chih and Tsu Keng-Chih was able to calculate π to *ten* decimal places; the details of their methodology have been lost. But we have their result, and it is correct.

As we have discussed elsewhere in this book, mathematicians of the Renaissance tended to be rather secretive about their work. Many of these scholars worked in isolation and not at universities. There was a considerable amount of professional jealousy, and little motivation to publish or to share ideas. It was not at all uncommon for a scholar to announce his results, but not reveal his methods. Some talented investigators, such as Pierre de Fermat, did so out of a sense of puckishness. Fermat would prove his theorems and then pose them as challenges to his fellow mathematicians. Some scholars did it because they did not trust other scientists. Henry Oldenburg pioneered the idea of the refereed scientific journal in 1665, and that really changed the tenor of scientific communication.

In the nineteenth century, at least among European mathematicians, communication blossomed. Weierstrass, Cauchy, Fermat, and many others conducted copious correspondence with scholars at other universities. Many of these letters survive to this day and are fascinating to read. In fact, Gösta Mittag-Leffler's rather palatial home in Djursholm, Sweden, is now a mathematics institute. Conferences and meetings and long-term workshops are held there regularly. What is remarkable about this particular math institute is that it still appears very much as it did when Mittag-Leffler lived in it. Much of his furniture survives, his library is intact, his photo albums and scrapbooks still sit on the shelf, and the papers in his study are all intact. Residing on a particular shelf in the study are a box with letters from Cauchy, a box with letters from Weierstrass (Mittag-Leffler's teacher), and many others as well. This is a real treasure trove, and fodder for future historians of mathematics.

In the twentieth century the habit of regular correspondence among mathematicians continued to develop. But something new was added. In the early part of the twentieth century there were few formal mathematics journals and few mathematical book publishers. So it became quite common for people to circulate sets of notes. This came about because a professor would give a course at the university developing some new ideas. The consensus often was that these ideas were valuable and should be promulgated. So a secretary was enlisted to type up the notes. Note that photocopying did not exist yet (there *were* more primitive methods of making copies). Sometimes just a few carbon copies of the notes were produced. Other times they were typed up on a ditto or mimeo master so that many copies could be made. In many (but not all) cases one could request a copy simply by sending in a modest amount of money to cover copying costs and postage.

These sets of notes were one of the primary tools of mathematical communication in the 1920s, 1930s, 1940s, and 1950s. One thing that is notable about the Princeton University mathematics library is that it has a special room with a magnificent collection of sets of notes from all over the world. Many of these are priceless, and contain extremely valuable information and ideas that are not reproduced elsewhere.

On October 4, 1957, the Soviets launched their unmanned satellite Sputnik. This event caught the Americans by total surprise; they suddenly realized that they were behind in the "space race," and behind in the development of technology in general. The government launched a huge push to accelerate America's high-tech growth, and this included a major development in education. As a result, American universities in the 1960s enjoyed an enormous mushrooming and development. Many new campuses were built, and existing campuses were expanded at a breakneck pace. And there were ripple effects. Mathematical publishing, in particular, really took off. A great many mathematical publishing houses sprang forth, and produced a host of mathematics books. Many new journals were begun. Several publishers began special series called "Lecture Notes," and they then published informal manuscripts that formerly would have been circulated privately as mimeographed sets of notes. Mathematical communication truly grew and prospered during this period.

Electronic communication today has changed everything again—probably for the better. In our modern world people routinely post sets of notes—created in TeX (see Section 9.2) so that they are typeset to a very high standard—on Web sites. People post their papers, as soon as they are written, on preprint servers. These are Web sites equipped with a mechanism that makes it easy for the user to `ftp` or upload a TeX or `*.dvi` or `*.PDF` file and then have it posted—for all the world to see—automatically and easily. Furthermore, the end user can download the paper, print it out, edit it, and manipulate it in other ways: send it on to others, share it with collaborators, or cut it up and combine it with other documents (although this latter is not encouraged).

Mathematicians can also send their papers directly to selected individuals using email; the device of the email *attachment* is particularly useful for this purpose. It should be noted here that scientific communication over email takes two forms: (a) one can send a formally written paper as an email attachment and (b) one can send remarks, or nascent ideas, as part of an email. The latter can be rather informal, and may contain opinion and even personal remarks.

A special feature of email is that it can affect the *quality* of communication. An email exchange can escalate in intensity rather rapidly. We have all had the experience of having a polite email exchange become ever more heated until the participants are virtually shouting at each other over the wires. The trouble with electronic communication is that you don't have the opportunity to look your interlocutor in the eye; you cannot use body language; you do not have room to interpret and to allow for nuance. Email is too hard-edged.

And this feature has consequences for scientific communication as well. An interesting story, coming to us from Harvard mathematician Arthur Jaffe [JAF], illustrates the point:

> In 1982, Simon Donaldson, then finishing his doctoral thesis at Oxford, advocated the study of the space of solutions to the Yang–Mills [field equations] on 4-manifolds as a way to define new invariants of these manifolds. Literally hundreds of papers had been written after Donaldson's initial work in 1983, and an industry of techniques and results developed to study related problems . . .
>
> This mathematical focus changed overnight in 1994, following a suggestion by Seiberg and Witten that a simpler approach might be possible. I heard this as a concluding remark in a physics seminar on October 6, 1994, held at MIT . . .
>
> So after that physics seminar on October 6, some Harvard and MIT mathematicians who attended the lecture communicated the remark by electronic mail to their friends in Oxford, California, and other places. Answers soon began to emerge at breakneck speed. Mathematicians in many different centers gained knowledge and lost their sleep. They re-proved Donaldson's major theorems and established new results almost every day and every night. As the work progressed, stories circulated about how young mathematicians, fearful of the collapse of their careers, would stay up night after night in order to announce their latest achievement electronically, perhaps an hour—or even a few minutes—before some competing mathematician elsewhere. This was a race for priority, where sleep and sanity were sacrificed in order to try to keep on top of the deluge of results pouring in. Basically ten years of Donaldson theory were reestablished, revised, and extended during the last three weeks of October, 1994.

Electronic communication—email and the World Wide Web—have transformed the development of mathematics in marvelous and profound ways. In fact it has changed the nature of our discourse. For papers that are sent around by email or posted on a preprint server are not published in the usual fashion.[85] They are not vetted, refereed, or reviewed in any manner. There is no filter for quality or appositeness or correctness. There is nobody

---

[85]They could eventually be published in a refereed journal, but at first they are not.

to check for plagiarism, or calumny, or libel, or slander.[86] There is just an undifferentiated flood of information and pseudoinformation.

And the end user must then somehow figure out what is worth reading and what is not. A traditional journal of high quality conducts a serious filtering operation. It is in its best interest to maintain a high standard and to showcase only the best work. This keeps their clientele coming back for more, and willing to pay the rather high tariff for some modern scientific journals.[87] The scientist looking for literature is likely to gravitate to names that he/she knows and trusts. It is quite easy to do a `Google` search for a particular author and to find all his/her works. Likely as not, they will all be posted on his/her Web site, or on a preprint server, and easily downloaded and printed out. But such a system does not serve the *tyros* well. A mathematician or scientist, who is just starting out will not be well known, so it is rather unlikely that some bigwig at Harvard is going to do a Web search looking for his/her work.

For many mathematicians, posting work on the Web has replaced dealing with traditional publishers who purvey traditional journals. When you post on the Web, you don't need to deal with *pompous* editors and *surly* referees. You just post and then go on about your business. You get feedback from interested readers, but that will be mostly benign and welcome. The point here is that the Web is magnificent for disseminating work broadly, and virtually for free. It is not good for archiving, since nobody knows how to archive electronic media. Each copy of an electronic product is unstable, and sunspots, or cosmic rays, or an electromagnetic storm could wipe out every copy on earth in an instant. A new, insidious virus could destroy 95% of the world's computers simultaneously. Mirror sites, disaster backups, the modified-Tower-of-Hanoi protocol, and other devices make our electronic media somewhat secure. But there is a long way to go. Each copy of a hard copy journal is stable: Print a thousand hard copies of the new issue of your journal and distribute them to libraries all over the world, and you can be reasonably confident that at least some of these will survive for several hundred years.

And while there are electronic journals that conduct reviewing and refereeing in the traditional manner, and to a very high standard, they are in the minority. Many electronic journals are a free-for-all, with the expected consequences and side effects. And there are also side effects on paper journals. Editors and referees, taken as a whole, are overworked. There are far too many papers.[88]

Electronic dissemination of scientific work raises a number of important side issues. We again borrow from Arthur Jaffe [JAF] as we lay out what some of these are (see also the book [KRA3] for useful information):

- How can we maintain standards of quality when the volume of publication is theoretically unlimited? For example, how can we continue to ensure serious refereeing, in spite of an enormous increase of volume brought about by the ease of word processing? This

---

[86]But `Google` *can* be used to check for plagiarism, and teachers commonly do so.

[87]This could be as high as $500 or $1000 per annum or much higher. One brain science journal these days costs $23,000 per year.

[88]The vehicle *Math Reviews* of the American Mathematical Society—which archives and reviews the majority of new math papers each year—reviews more than 75,000 papers per annum.

phenomenon will overload a system already near the breaking point. In fact, there are already so many publications that careful refereeing is hard to find.

- How can we organize publications so we can find what we need? Casual browsing of the Web illustrates this problem.
- How can we establish the priority of ideas and assign credit? Volume makes this question extremely difficult, as do multiple versions of the same work. The possibility of interactive comments makes it even more difficult. Will some people resort to secrecy to ensure priority, while others will make exaggerated claims? Will much work get lost because it was not fashionable at the time?
- How can we prevent fads from overwhelming groundbreaking long-term investigation? To what extent will frequency of citation be taken as a gauge of success?
- How can we archive our work? In other words, how can we ensure access to today's publications in 5, 10, 30, or in 100 or 300 years? Libraries have learned to live with paper, and the written word is generally readable. But technology develops rapidly. Formats, languages, and operating systems change. We know that the American Mathematical Society has a roomful of 10-year-old computer tape, unusable because the computer operating system has changed. How can we be assured that our mathematical culture will not disappear because a change in programs, a change in media, or another change in technology makes it too expensive to access a small number of older works?

9

Aspects of Modern Mathematical Life

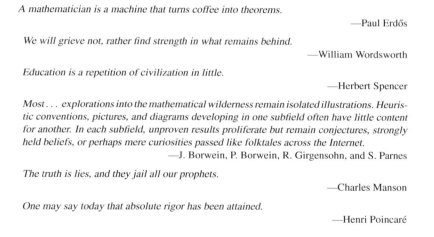

A mathematician is a machine that turns coffee into theorems.

—Paul Erdős

We will grieve not, rather find strength in what remains behind.

—William Wordsworth

Education is a repetition of civilization in little.

—Herbert Spencer

Most... explorations into the mathematical wilderness remain isolated illustrations. Heuristic conventions, pictures, and diagrams developing in one subfield often have little content for another. In each subfield, unproven results proliferate but remain conjectures, strongly held beliefs, or perhaps mere curiosities passed like folktales across the Internet.

—J. Borwein, P. Borwein, R. Girgensohn, and S. Parnes

The truth is lies, and they jail all our prophets.

—Charles Manson

One may say today that absolute rigor has been attained.

—Henri Poincaré

9.1   The World We Live In

As this book has endeavored to describe, prior to two hundred years ago the mathematician was of necessity a dedicated, independent, largely self-sufficient lone wolf. There were few academic positions and no granting agencies (of the government or otherwise). If a mathematician was extraordinarily lucky, a patron could be found who would provide financial and other support. Euler and Descartes had patrons. Galois and Riemann did not.

Today's mathematical life is quite different. There are thousands of colleges and universities all over the world—more than 4100 in this country alone. Many government agences and laboratories employ mathematicians. Many private firms—from General Electric to Microsoft to Aerospace Corporation—employ mathematicians. A number of mathematicians are consultants, and many others work in the publishing world.

There are many aspects to the mathematical life in the modern world. Some mathematicians, especially those at the university, study mathematical theory and prove theorems. Some, especially (but not exclusively) those in industry, develop applications of mathematics. Some ply their trade by developing computer products that implement mathematical ideas. Some work in the financial sector, doing financial forecasting and developing investment strategies. Some work as actuaries, producing mortality tables and designing retirement plans.

A mathematician can be supported (financially) by a university or college, by a government agency, by a commercial employer, by one or more grants, by a financial benefactor, or by some combination of these. Most mathematicians make a pretty good living, and some are doing quite well.

There are mathematicians who served as technical advisors for the film *A Beautiful Mind*, others who serve as technical advisors for the television show *Numbers*, and still others who work with *Pixar* and other high-tech firms to create special effects for Hollywood.

In the old days mathematics was a cottage industry. Today it is a vital part of life, integrated into every aspect of human inquiry. The mathematical life is hard work, but it is rewarding and meaningful. It has many aspects and offers many opportunities.

9.2 Mathematics Institutes

Most mathematics institutes are funded with government money. The mathematics institute in Bures-sur-Yvette and in Luminy in France, the Isaac Newton Institute in Cambridge, England, the Banff International Research Station in Banff, Canada, the famous institute IMPA (Instituto Nacional de Matemática Pura e Aplicada) in Brazil, the Mathematisches Forschungsinstitut in Oberwolfach, Germany are all funded with national government monies. There are eight federally funded mathematics institutes in the United States. But it is remarkable that the United States enjoys a number of privately funded mathematics institutes as well.

The first private mathematics institute in this country was founded in the 1930s by Louis Bamberger of Bamberger's Department Stores. Working with the prominent academician Abraham Flexner, Bamberger was ultimately convinced to establish an "Institute of Useless Knowledge." And thus was born the Institute for Advanced Study in Princeton, New Jersey. Centered at Fuld Hall on lovely grounds that are also a game preserve, the Institute is a haven for very advanced research activity in mathematics, astrophysics, history, social science, and other areas. When Flexner and his collaborators were setting up the Institute for Advanced Study, one of their first jobs was to select the founding faculty. These faculty would be permanent members, would be paid princely salaries, and would be charting an important course for intellectual development in this country. So the selection had to be made with utmost care. What they found was that the only subject area in which they could obtain a real consensus on who were the best people was in mathematics (and mathematical physics). So that is where they started. And one of their very first appointments was Albert Einstein. This appointment that really put the fledgling Institute for Advanced Study on the map. Over the years, the Institute for Advanced Study has had an astonishing panorama of powerful and influential faculty. It has hosted any number of important conferences and intellectual events. On the whole, it has been a great success as a scholarly institution.

John Fry was a college student at the University of Santa Clara. There he fell under the spell of mathematics—particularly under the tutelage of Professor Gerald B. Alexanderson. Fry was a football star in college, and also was a good friend of math major J. Brian Conrey. Well, Fry went on to become a successful retailer. Specifically, his father owned a chain of supermarkets. At some stage he gave each of his children $1 million to see whether they could make something of themselves. John Fry built the very successful *Fry's Electronics* chain of emporium-style electronics stores. There are now 33 of them in four states; there are plans to expand across the country and around the world, and John Fry is worth over $1 billion. In spite of his success and wealth (or perhaps because of it), Fry has maintained his interest in mathematics. He has been collecting rare mathematics books for some years. Included in his collection are Napier's book on logarithms, some of Euler's early books, and a first edition of Isaac Newton's *Principia* signed by Stephen Hawking (who presently holds Newton's Lucasian chair at Cambridge University).

In 1994 John Fry decided to start a mathematics institute. Called The American Institute of Mathematics (AIM), it is presently located in Palo Alto, California, in the former corporate headquarters of Fry's Electronics. AIM enjoys generous funding each year from John Fry, and it also has funding from the federal National Science Foundation (the Princeton Institute for Advanced Study has similar support from NSF). The Director of AIM is Brian Conrey, Fry's old college friend.

AIM has purchased a 1900-acre plot in Morgan Hill, California, (just southeast of Silicon Valley), and is building a world-class home for its mathematics institute. It is modeled, in considerable detail, after the medieval Moorish fortress known as the Alhambra; but it will be about three times the size. The grounds are also home to a beautiful 18-hole custom golf course where John Fry enjoys an occasional game of golf with his friends. Fry has also hired two famous chefs—who formerly worked at fancy resorts in northern California. They are currently available to cook the occasional lunch for Fry and his golfing buddies. But when the new institute is up and running in Morgan Hill, these two will prepare the meals for the hard-working mathematicians.

A new addition to the "privately funded mathematics institute" scene is the Clay Mathematics Institute in Cambridge, Massachusetts. Founded in 1998 with money from Landon T. Clay (a venture capitalist who is a direct descendent of Cassius Clay, the Civil War general), this institute has really made a splash. The things you can do with $50 million. And one of the remarkable things that the Clay Institute has done is to establish the "Millennium Problems." This is a collection of seven well-known mathematical problems that have been dubbed the most prominent, important, and challenging unsolved problems in all of mathematics.[89] These are *the* mathematics problems for the 21st century. What is remarkable is that the Clay Institute has offered a $1 million prize for the solution of any of these seven problems.

---

[89] A detailed and authoritative discussion of these problems appears in [CJW]. See also [DEV1]. This situation is quite similar to what Hilbert did in 1900. But there are two important differences. One is that this time a *committee* of distinguished scientists put together the list of problems. The other is that each of these new problems has a $1 million bounty attached to it. Only the Riemann hypothesis is both on Hilbert's list and on the Clay list.

A typewriter, by nature, is a *monospaced* environment. This means that all the characters are the same size and width, with the same amount of space between them. Vertical spacing is quite tricky. The result of the IBM Selectric rendition of a paper was readable mathematics, but not the way it is typeset in the mind of God.

But in the early 1980s everything changed dramatically. Donald Knuth of Stanford University invented the computer typesetting system TeX. This book is typeset in TeX. In fact today most mathematics is typeset in TeX. TeX is a high-level computing language like Fortran or Java. It is *not* a word processor. When you create a document in TeX, you are in fact issuing commands (accurate to within a millionth of an inch) as to how you want each character and each word to appear and to be positioned on the page. Some have called the invention of TeX the most important event in the history of typesetting since Gutenberg's printing press.

The package LaTeX was developed by Leslie Lamport in 1983 to make TeX easier to use. Lamport provides useful macros to make the TeX commands—especially formatting commands—simpler and more intuitive. To give the reader an idea of how TeX and LaTeX work, we now provide a sample. The user opens up a new file and, using a text editor, inputs LaTeX code that looks something like this:

```
********************************************************
\documentclass{article}

\newfam\msbfam
\font\tenmsb=msbm10    \textfont\msbfam=\tenmsb
\font\sevenmsb=msbm7   \scriptfont\msbfam=\sevenmsb
\font\fivemsb=msbm5    \scriptscriptfont\msbfam=\fivemsb
\def\Bbb{\fam\msbfam   \tenmsb}

\def\RR{{\Bbb R}}
\def\CC{{\Bbb C}}
\def\QQ{{\Bbb Q}}
\def\NN{{\Bbb N}}
\def\ZZ{{\Bbb Z}}

\begin{document}

Let $f$ be a function that is continuous from complex numbers
$\CC$ to itself.  Consider the
auxiliary function
$$
g(z) = \frac{\int_a^b \frac{\alpha z + \beta}{\gamma z
            + \delta} \, dz}{[\cos z]^z \cdot z^{\sin z}} \, .
$$
Then
$$
g \circ f (z)
$$
```

```
operates in a natural manner on the Banach space of continuous
functions on the unit interval $I$.

\end{document}
```

****

The user compiles this LaTeX source code using a standard LaTeX compiler that is available commercially, or by download from the Web. The output looks like this:

****

Let f be a function that is continuous from the complex numbers $\mathbb{C}$ to itself. Consider the auxiliary function

$$g(z) = \frac{\int_a^b \frac{\alpha z + \beta}{\gamma z + \delta}\, dz}{[\cos z]^z \cdot z^{\sin z}}.$$

Then

$$g \circ f(z)$$

operates in a natural manner on the Banach space of continuous functions on the unit interval I.

****

The user may send this output to a screen, to a printer, to a FAX machine, to the Internet, to a `*.PDF` translator, to a `*.ps` translator, as an attachment to an email, or to a number of other devices. TeX is a powerful and flexible tool that has become part of the working life of every mathematician.

Today most mathematicians know how to use TeX, and they type their own papers in TeX. The result is that most mathematics papers now are actually typeset, as they would be in a finished book. Mathematical characters are sized and positioned and displayed just as they would be in a typeset script for a first-class mathematical monograph. This opens up many possibilities.

A paper that is prepared in TeX is easily exported to `PostScript` or an Acrobat-readable format (such as `*.PDF`). Such a file in turn may easily be posted on the World Wide Web. A paper in `*.PDF` that is posted on the Web will have all mathematical formulas, all text, and all graphics (i.e., figures) displayed as they would be in a published paper or book. And many mathematicians choose to disseminate their work in just this fashion as soon as it is completed. This is what electronic preprint servers are for. A preprint server is a computer hooked into the World Wide Web that serves as a forum for mathematical work. It archives thousands of papers in an attractive and accessible fashion, offers the end user a choice of formats, and displays each paper as it would appear in hard copy—with all the mathematical notation, all the text, all the graphics, all the formulas.

A preprint server is a clearinghouse for new mathematical work. It makes the fruits of our labors available for free to everyone in the world who has access to a computer. For rapid and free dissemination of new scientific ideas, there is nothing to beat a preprint server.

But it should be stressed that an electronic preprint server engages in *no vetting* of the work that it displays. There is no refereeing and no reviewing. And this is as it should be. For the people running the preprint server can assume no responsibility for the work that it displays (if they conducted reviewing, then they *would* bear some legal as well as ethical responsibility). Moreover, it is most desirable that the preprint server should run without human intervention. Anyone at all should be able to post an article on the server with no assistance or manipulation from the people who run the server.

The most prominent and influential and important mathematics preprint server in the world today is arXiv. Begun at Los Alamos by Paul Ginsparg—over the objections and distinct lack of support of his supervisors—arXiv began as a rather small-scale preprint server in high-energy physics and is now a very large enterprise that encompasses physics, mathematics, and other subjects as well. There are tens of thousands of preprints now available on arXiv for free and easy download to anyone who wishes to read the work. And the number is growing by leaps and bounds with each passing day. Ginsparg has been awarded a MacArthur Fellowship for his work in developing arXiv. He has also moved to Cornell University, where he is a Professor of Physics. And now arXiv makes its home at Cornell. A happy ending for all.

10

Beyond Computers: The Sociology of Mathematical Proof

It's one of those problems [the Kepler sphere-packing problem] that tells us that we are not as smart as we think we are.

—D.J. Muder

The packing will be the tightest possible, so that in no other arrangement could more pellets be stuffed into the same container.

—J. Kepler

It became dramatically clear how much proofs depend on the audience. We prove things in a social context and address them to a certain audience.

—William P. Thurston

Whether the turn of the century will find many open questions under attack by a team consisting of a researcher and an automated reasoning program is left to time to settle. Clearly, the preceding decade has witnessed a sharp increase in this regard, sometimes culminating in the answer to a question that had remained open for decades. Thus we have evidence that an automated reasoning program can and occasionally does contribute to mathematics and logic.

—Keith Devlin

All truths are easy to understand once they are discovered; the point is to discover them.
—Galileo Galilei

Mathematicians are like a certain type of Frenchman: when you talk to them they translate it into their own language and then it soon turns into something entirely different.
—Johann Wolfgang von Goethe

10.1 The Classification of the Finite Simple Groups

The idea of "group" came about in the early nineteenth century. A product of the work of Évariste Galois (1812–1832) and Augustin-Louis Cauchy, this was one of the first cornerstones of what we now think of as *abstract algebra*. So what is a group?

The idea is simplicity itself. A *group* is a set (or a collection of objects) G equipped with a binary operation that satisfies three axioms. Now we are seeing mathematics in action. In particular, we are encountering definitions and axioms.

What is a "binary operation"? This is a way of combining two elements of G to produce another element of G. For instance, if our set is the whole numbers (or integers), then ordinary addition is a binary operation. For addition gives us a way to combine two whole numbers in order to produce another whole number. As an example,

$$2 + 3 = 5.$$

We combine 2 and 3 to produce 5. Multiplication is another binary operation. An example of multiplication is

$$2 \times 3 = 6.$$

We combine 2 and 3 to produce 6.

These last two examples are rather pedestrian. What is a more exotic example of a binary operation? Consider 2×2 matrices. We multiply two such matrices according to this rule:

$$\begin{pmatrix} a & b \\ c & d \end{pmatrix} \cdot \begin{pmatrix} \alpha & \beta \\ \gamma & \delta \end{pmatrix} = \begin{pmatrix} a\alpha + b\gamma & a\beta + b\delta \\ c\alpha + d\gamma & c\beta + d\delta \end{pmatrix}.$$

We see that this gives a way of combining two 2×2 matrices to give another 2×2 matrix.

So the group G has a binary operation, as described above, which we denote by $\cdot$ (we do this as a formal convention, whether the operation in any specific instance may turn out to be addition or multiplication or some other mode of combining elements). And the axioms that we put in place that govern the behavior of this binary operation are these:

(1) The binary operation is associative:

$$a \cdot (b \cdot c) = (a \cdot b) \cdot c.$$

(2) There is a special element $e \in G$, called the *identity element of the group*, such that $e \cdot g = g \cdot e = g$ for every $g \in G$.

(3) Each element $g \in G$ has an *inverse element*, called g^{-1}, such that

$$g \cdot g^{-1} = g^{-1} \cdot g = e.$$

It turns out that groups arise in all aspects of mathematics, physics, and even engineering. Some examples are these:

- *The integers* (i.e., the positive and negative whole numbers), equipped with the binary operation of addition, form a group. Integer addition is associative: We know that $a + (b + c) = (a + b) + c$. The identity element e is just 0, for $0 + a = a + 0 = a$ for all elements a in the integers. And finally the additive inverse for any integer a is $-a$.

- *The positive real numbers* (i.e., all positive whole numbers, all positive rational numbers, and all positive irrational numbers), equipped with the binary operation of multiplication, form a group. In this context associativity is a standard arithmetical fact. The identity element is just 1. And the multiplicative inverse of a is $1/a$.

- *The 2×2 matrices* having nonvanishing determinant, so that

$$\det \begin{pmatrix} a & b \\ c & d \end{pmatrix} = ad - bc \neq 0,$$

and equipped with the binary operation of matrix multiplication (defined above), form a group. Associativity of matrix multiplication is a standard fact from linear algebra. The identity element is the matrix $\begin{pmatrix} 1 & 0 \\ 0 & 1 \end{pmatrix}$. And the inverse of a given matrix A is its matrix multiplicative inverse, which one can calculate to be

$$A^{-1} = \begin{pmatrix} \dfrac{d}{ad - bc} & \dfrac{-b}{ad - bc} \\ \dfrac{-c}{ad - bc} & \dfrac{a}{ad - bc} \end{pmatrix}.$$

We invite the reader to perform the simple matrix multiplication and verify that

$$A^{-1} \cdot A = A \cdot A^{-1} = \begin{pmatrix} 1 & 0 \\ 0 & 1 \end{pmatrix}.$$

- *Group theory* is used to describe the algebra of bounded operators on a Hilbert space. These in turn, according to a profound idea of Werner Heisenberg (1901–1976), Erwin Schrödinger (1887–1961), and John von Neumann, can be used to explain the structure of quantum mechanics.

- Every cell phone is encoded with a copy of the Cayley numbers. This is a special group that finds applications in mathematical physics, which is used to encrypt the information being transmitted by a cell phone.

Since groups are such universal mathematical objects, and since they can be used to describe, or control, or analyze so many different types of physical phenomena, there is great interest in classifying all the groups that may arise. Great progress has been made in this regard with respect to the *finite simple groups*. A group is finite if it has just finitely many elements. An elementary example of a finite group is this. Consider the polygon (i.e., the *square*) exhibited in Figure 10.1. It has certain symmetries: (i) We can flip it left-to-right, (ii) We can flip it top-to-bottom, (iii) We can flip it along the main diagonal, (iv) We can flip it along the minor diagonal, (v) We can rotate it through 90° or 180° or 270°. The collection of all these symmetries (see Figure 10.2) forms a group—where the binary operation is composition (i.e., superposition) of operations. And it is clear that this group has just finitely many elements.

The concept of "simple group" is somewhat more technical. A group is *simple* if it cannot be decomposed into more elementary pieces, each of which is a group. It is a fundamental structure theorem that *any* finite group is made up of simple groups. So the problem of classifying all finite groups reduces, in some sense, to classifying all the finite simple groups.

The classification of the finite simple groups is one of the great triumphs of twentieth century mathematics. What is interesting about this achievement—from our point of view—is that this mathematical result is not the work of any one scientist. It is also

Figure 10.1. The square.

not the work of two collaborators, or of a small team working together at a mathematics institute. The classification of the finite simple groups follows from the aggregate of the work of hundreds of mathematicians in dozens of countries stretching back to the middle of the nineteenth century.

The classification of finite simple groups comprises more than 10,000 pages of dense mathematical scholarly writing. There exists no published outline of the proof.[92] The proof has existed in some form since 1981. Gaps have been found along the way.[93] Those that are known have now been fixed.

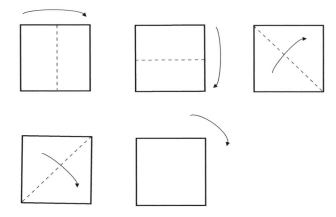

Figure 10.2. The symmetries of the square.

---

[92]The book [GLS] contains something that resembles an outline of the classification of the finite simple groups as it stood at that time (1994). The proof has evolved since then.

[93]There are gaps in mathematics and there are *gaps*. The biggest gap in the finite simple group program was to classify the "quasithin" groups. This task had been assigned to a young assistant professor at a West Coast university. This is what the fellow had studied in his thesis. But it turned out that there were significant *lacunae* in his arguments. These have recently been filled in and rectified by experts Michael Aschbacher and Stephen Smith (see [ASM]). Their two volumes, written to address the questions about the quasithin groups, comprise 1200 pages!

The nearest thing that we have to a "record" of the classification of the finite simple groups is the four-volume work *The Classification of the Finite Simple Groups* by Gorenstein, Lyons, and Solomon. It totals 2139 pages of dense mathematical argumentation, and gives a substantial idea of what the program is all about. But Michael Aschbacher, a leading authority in the field, has estimated recently that the *full proof* will be more than 10,000 pages. And bear in mind that that is 10,000 pages of published mathematics in the standard modern argot, which means that it is pithy, condensed, and brief, and leaves a fair amount of work to the reader.

It was Daniel Gorenstein (1923–1992) of Rutgers University who, in the early 1970s, convinced people that the holy grail was in sight. He assembled a conference of over 100 experts in the field and took it upon himself to organize them into an army that would attack the considerable task of nailing down this proof. He summarized the state of the art and pointed out what was accomplished and what needed to be done, and to actually convince people to fill in the gaps and prove the results that were still outstanding. He assigned specific tasks to individuals and teams from all over the world. To quote Michael Aschbacher [ASC]:

> ... Danny Gorenstein began to speculate on a global strategy for a proof. In effect he called attention to certain subproblems, which appeared to be approachable, or almost approachable, and he put forward a somewhat vague vision of how to attack some of the subproblems, and how his various modules might be assembled into a proof. While his program was sometimes a bit far from what eventually emerged, in other instances he was fairly prescient. In any event, Gorenstein focused on the problem of classifying the finite simple groups, in the process making the effort more visible. He also gave it some structure and served as a clearing house for what had been done, and was being done. In short, Gorenstein managed the community of finite simple group theorists, and to a lesser extent, managed part of the development of the proof itself.

It finally got to the point, in the late 1970s that there was just one remaining hole in the program. Bernd Fischer and Robert Griess had predicted in 1973 that there was a "largest" sporadic simple group that became known as the *monster*. It was suspected to have

$$808017424794512875886459904961710757005754368000000000$$

elements, but no one could in fact identify this group—people were not even sure of its size. Researchers had computed its character table, long before they found the group. So a lot was known about the mysterious monster. *Except* whether it actually existed. The group held a special fascination, because the prime factorization

$$808017424794512875886459904961710757005754368000000000$$
$$= 2^{46} \cdot 3^{20} \cdot 5^9 \cdot 7^6 \cdot 11^2 \cdot 13^3 \cdot 17 \cdot 19 \cdot 23 \cdot 29 \cdot 31 \cdot 41 \cdot 47 \cdot 59 \cdot 71$$

also exhibits the atomic weights of the factors in an important molecule. In 1980, Griess was able to construct the monster group as the automorphism group of the Griess algebra (in fact the group can be generated by two 196882×196882 matrices over the group with two elements, and John Horton Conway has simplified the construction even further).

The uniqueness of this special group was established by Griess, Meierfrankenfeld, and Segev in 1990. And that put the cherry on the top of the program. The classification of the finite simple groups was (in principle) complete!

The next step, naturally, was for Daniel Gorenstein to arrange for some books to be published that would archive the entire proof, going back to the very beginning in the early nineteenth century and running up to the present day when the final profound ideas were put into place. It was estimated that this work would comprise several volumes of total length in the (many) thousands of pages. To date, this comprehensive work on the classification of the finite simple groups has not been completed.

Even today, the experts admit that the status and validity of this immense proof is still in doubt. Expert Michael Aschbacher [ASC] says:

> First, the proof of the Classification is very long and complicated. As a guess, the proof involves perhaps 10,000 pages in hundreds of papers, written by hundreds of mathematicians. It would be difficult to establish exactly which papers are actually a necessary part of the proof, and I know of no published outline. At least this last difficulty will be eliminated by a program in progress, whose aim is to carefully write down in one place a complete and somewhat simplified version of most of the proof. Still there has not been as much improvement and simplification of the original proof as one might expect. ... By consensus of the community of group theorists, the Classification has been accepted as a theorem for roughly 25 years, despite the fact that, for at least part of that period, gaps in the proof were known to exist. At this point in time, all known gaps have been filled. The most significant of these (involving the so-called 'quasithin groups') was only recently removed in the lengthy two volume work of Aschbacher and Smith. During the 25 years, the proof of the Classification has not evolved as much as one might expect. Some simplifications and conceptual improvements to certain parts of the argument have emerged, and there is a program in progress to write down the proof more carefully in one place. Dependence on computer aided proofs for the existence and uniqueness of the so-called sporadic groups has been almost entirely eliminated. But for the most part the proof still has the same shape and complexity.

But now the experts in finite simple group theory are working on a "second generation" proof, one that will be more accessible, more coherent, and (one hopes) rather shorter. This mathematical achievement is important to document no matter how many pages it takes, but it is more likely that people will be able to read it, understand it, verify it, and internalize it if it is briefer and more attractively written.

This really is a grand saga, and a tribute to the spirit of cooperation that is prevalent among modern mathematicians. But there have been some stumbles and hiccups along the way. We have noted that there was one mathematician at a west coast university who was assigned a certain part of the program. He labored away at this for years, produced a manuscript of several hundred pages, but ultimately became discouraged and never finished his task. Unfortunately Gorenstein died and the rest of the world lost track of this little piece of the puzzle. On those rare occasions when, at a conference, somebody said, "But what about that part—the quasithin stuff—that so-and-so is supposed to be doing?", the typical answer was, "Don't worry. The program is so robust that it's bound to work out. We have more important concerns at this time." It was finally Fields Medalist Jean-Pierre Serre who wrote a paper pointing out this nontrivial gap in the literature—a gap that had to be filled. Serre's article caused quite a stir, and in the end Aschbacher and Smith [ASM] had to work out the ideas and write them up. This took two volumes and 1200 pages. And that was to fill just one gap.

Of course, as we have indicated, there are experts all over the world who spend virtually all their time studying the classification of the finite simple groups and searching for ways to simplify and clarify the arguments. And it does happen that every now and then, someone discovers a notable glitch in the big picture. But the program does appear to be stable, the glitch always gets fixed, and there is every reason to believe that eventually there will exist a multivolume work documenting this theorem that took nearly two centuries to prove.

There are still many more volumes to be written in the saga of the finite simple groups. Part of the problem is that this is a human endeavor, and a great many of the key participants are aging or retiring or dying or some combination of these. It is not a sure bet that Gorenstein's vision will be carried out and validated as he originally anticipated.

10.2 Louis de Branges's Proof of the Bieberbach Conjecture

The Bieberbach conjecture is one of the grand old questions in complex variable theory. It concerns the nature of the power series coefficients of certain types of analytic functions on the unit disk in the complex plane. More precisely, we have an analytic function

$$f(z) = z + \sum_{j=2}^{\infty} a_j z^j$$

defined on the unit disk $D = \{z \in \mathbb{C} : |z| < 1\}$. We assume that the function takes distinct domain points z_1, z_2 to distinct targets. The conjecture is that it must be that the jth coefficient a_j cannot be any larger (in modulus or absolute size) than j.

The full details need not concern us here. Suffice it to say that this is a rather technical mathematical question that only a specialist would care about, and Louis de Branges of Purdue University cared about it passionately.

Louis had the reputation of being a talented mathematician with a lot of good ideas. But he also had the reputation of something of a crank, because he had cried "wolf" once too often. It should be stressed that all mathematicians make mistakes. Any first-rate mathematician is going to take some risks, work on some hard problems, shoot for the moon. In doing so, this person may become convinced that he/she has solved a major problem. And thus mistakes get made. Almost any good mathematician has published a paper with mistakes in it. Many of us have published papers that are just plain wrong. And the referee did not catch the mistake either, so it must have been a pretty good mistake.

There is another famous problem in mathematical analysis called the "invariant subspace problem." A great many people would love to solve this problem—about the nature of bounded operators on a Hilbert space (this is the language in which the modern theory of quantum mechanics is formulated). But nobody has succeeded. Unfortunately, Louis had

---

Complex Variables

A good part of the history of mathematics was concerned with the solving of polynomial equations. Given a polynomial $p(x) = a_0 + a_1 x + a_2 x^2 + \cdots + a_k x^k$, one wants to find all values of x that make the polynomial vanish. Such a program is ill-fated, for the polynomial $p(x) = x^2 + 1$ has no real roots. Plug in any real x and you get a value that is at least 1.

The complex numbers were invented in large part to provide a universe in which all polynomials have roots. But it turns out that complex variables and complex functions are powerful devices for describing many phenomena in nature. Fluid dynamics, aerodynamics, geophysics, cosmology, and many other subjects lend themselves well to analysis with complex functions. Today every engineer must learn the theory of complex variables.

---

announced that he could do it, and he fell on his face. Yet another example is de Branges's claimed proof of the Ramanujan conjecture—later in fact proved by Fields Medalist Pierre Deligne.

So de Branges's credibility was somewhat in doubt in 1984 when he announced that he had a solution of the classic Bieberbach conjecture—a problem that went back for its original formulation to 1916. To further confound matters, de Branges claimed not that his proof was in a paper of 20 or 30 or 40 pages that one could just sit down and read. Rather, Louis de Branges declared that the proof was part and parcel of the new edition of his book *Hilbert Spaces of Entire Functions* [DEB1]. There was hardly anyone who was going to sit down and read a 326-page book in order to determine whether Louis had really done it. All other things being equal, Louis de Branges's "proof" of the Bieberbach conjecture could easily have moldered away in manuscript form for many years—with nobody having the time or interest to check it.

But Providence intervened on Louis's behalf! In the harsh winter of 1984, de Branges took a sabbatical leave, which he spent at the Steklov Mathematics Institute in St. Petersburg, Russia. Now the Russians have a strict and powerful mathematical tradition, and a great determination and work ethic. They spent the entire semester with de Branges distilling his proof of the Bieberbach conjecture from the 300+ page book manuscript. They found a number of gaps in de Brange's arguments and helped to fix them. These guys were heroes in the strongest sense of the word. And they got the job done.

The result is a 16-page paper [DEB2] that appears for anyone to check in *Acta Mathematica*. Now many hundreds of mathematicians have read de Branges's proof, and confirmed it. There is no doubt that the Bieberbach conjecture has been proved, and that Louis de Branges proved it (with the help of a terrific team of St. Petersburg mathematicians).

But there are even further developments. Louis de Branges engaged in the common mathematical practice of circulating copies of his paper long before its formal publication. This is called a *preprint*. It is a manuscript that is not handwritten—it is a *formal typescript*. It is an official enunciation of what this mathematician thinks has been proved, together with the proof itself. And Louis sent hundreds of these all over the world. Christian Pommerenke and Carl Fitzgerald were among the recipients of this largesse. And then an unfortunate accident occurred. Namely, Fitzgerald and Pommerenke found a number of interesting and decisive ways to simplify and clarify de Branges's arguments. Their ideas were sufficiently important to justify the publication of a separate paper. They wrote up their ideas and submitted them to the *Transactions of the American Mathematical Society* (see [FIP]). Through a variety of clerical SNAFUs, it came about that the very brief (only 8 pages) Fitzgerald/Pommerenke proof of the Bieberbach conjecture appeared in print *before* the original de Branges proof appeared.

Louis was not happy. Even though Fitzgerald and Pommerenke had been eminently respectful of Louis (they even used his name in their title, and not Bieberbach's!), there is no question that a major screw-up had occurred. At a celebratory conference the purpose of which was to hail Louis as the hero of the hour, Carl Fitzgerald introduced de Branges to speak. Professor de Branges stood up and announced to one and all that Fitzgerald and his collaborator were gangsters who were set to steal his ideas. None of it was very pretty.

And there are further developments. Lenard Weinstein wrote a thesis at Stanford University (his ideas are published in [WEI]) in which he dramatically simplified de Branges's proof. In fact his proof, which begins with the classical Loewner equation, uses nothing more than calculus. It is just four pages. Although not well known, this was a real milestone. But Doron Zeilberger [ZEI] took things a decisive step further. In their paper *A high-school algebra, "formal calculus," proof of the Bieberbach conjecture [after L. Weinstein]* [EKZ], Zeilberger and Shalosh B. Ekhad provide a proof of the Bieberbach conjecture that is just a few lines. It consists primarily of verifying a large combinatorial identity, and that verification can in fact be assigned to a computer. It is noteworthy that the original title of the Zeilberger/Ekhad paper was *A wallet-sized, high-school-level proof of the Bieberbach conjecture.* Journal editors, stodgy as they are, will not tolerate a paper with a whimsical title such as this. It could not stand. Instead the paper appears under the indicated title. The reader should find it particularly interesting to note that Shalosh B. Ekhad is *not* a person. In fact this is Zeilberger's nickname for his computer!

10.3 Wu-Yi Hsiang's Solution of the Kepler Sphere-Packing Problem

Together with his brother Wu-Chung Hsiang of Princeton University, Wu-Yi Hsiang of U. C. Berkeley established a preeminent reputation in the 1960s as an expert in the theory of group actions on topological spaces. This was a hot field at the time, and earned each of them jobs at the best mathematics departments in the world. At some point late in his career, Wu-Yi's interests began to wander, and the brothers settled on a fascinating problem that found its genesis with Sir Walter Raleigh. As we all know, Raleigh played a key role in the history of the United States. Among other things, he is responsible for our addiction to tobacco. But he was also an adventurer and something of a pirate. One day he questioned one of his gunners about the most efficient way to stack cannonballs. This question turned out to have mathematical interest, and was eventually communicated to the distinguished mathematician and astronomer Johannes Kepler. The classical formulation of Kepler's problem is then, "What is the most efficient way to pack balls of the same size into space?"

Kepler published this question in a booklet called *Strena sue de nive sexangula* in 1611. Interesting.

This is in fact a question that grocers face every day. If the produce person in a supermarket wants to display oranges (assuming that all the oranges are about the same size and shape), what is the most efficient means to stack them and display them? That is, how can we fit the most oranges into the least space? This may seem like a frivolous question, but it is not. And its solution is by no means obvious. The analogous question in two dimensions has had a ready solution for a long time, and it is well worth examining. Look first at Figure 10.3. It shows a number of 2-dimensional balls, or disks—all having the same radius—packed into the plane. These are displayed in the "rectilinear packing," which is one of the obvious ways to fit disks efficiently into a small space. It turns out that this is not the best way to do it—with the rectilinear packing, just 0.7854 of the plane is occupied by disks, and 0.2146 of the plane is not. A sophisticated

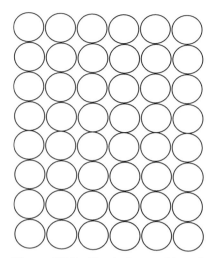

Figure 10.3   Not the best packing of disks.

exercise in matrix theory shows that the *best* way to pack 2-dimensional disks into the plane is the *hexagonal packing*, displayed in Figure 10.4. One can calculate that this is a distinctly better method—with the hexagonal packing, the disks occupy 0.9069 of space—and in fact show that it is the optimal method.[94] The Kepler sphere-packing problem is to find an analogous answer in three (or more) dimensions.

It is generally believed that the 3-dimensional analogue of the planar hexagonal packing is the optimal packing for spheres in three dimensions (refer to Figure 10.5). This is the "default" manner in which a grocer would stack oranges in a store display, or a gunner would stack his cannonballs. It has density $\alpha = 0.74048$. That is, with the described packing, 74.048 percent of space will be occupied by the cannonballs and 25.952 percent of space will be occupied by air. Old Mother Nature, with the way that she packs atoms into molecules, provides evidence for this belief. But there was, until recently, no proof.

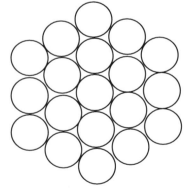

Figure 10.4   Best packing of disks in the plane.

---

[94] The history here is interesting. Carl Friedrich Gauss, about 200 years ago, showed that the hexagonal packing is the best among all possible *regular* packings. The Hungarian László Fejes Tóth proved in 1940 that the hexagonal packing is the best among *all possible* packings.

Kepler first formulated his problem in 1611. So this is one of the very oldest unsolved mathematics problems. Gauss contributed the first result on Kepler's problem. He showed that the Kepler conjecture is true if the spheres are assumed in advance to be arranged in a regular lattice. Thus, if there were to be a counterexample to the Kepler conjecture, then it would exhibit balls arranged in an irregular fashion. One of the most significant modern attacks on the problem was levied by László Fejes Tóth (see [FTO]). In 1953 he demonstrated that the problem could be reduced to a finite (but very large) number of calculations (using George Dantzig's idea of linear programming). Architect, geometer, and

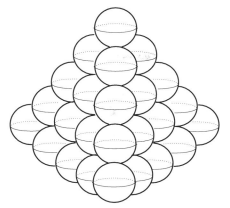

Figure 10.5 The 3-dimensional analogue of the planar hexagonal packing of disks.

entrepreneur Buckminster Fuller claimed in 1975 to have a proof of Kepler's conjecture, but this proved to be incorrect.

Interesting ideas about the problem in higher dimensions were offered by Conway of Princeton in [CON]. Wu-Yi Hsiang proposed to use a new implementation of spherical trigonometry—in fact he virtually reinvented the subject—to give a resolution of Kepler's problem in dimension three. In 1993 he wrote a long paper [HSI1] (92 pages), together with a secondary paper [HSI2] explaining what was going on in the first paper, laying out his solution of the Kepler sphere-packing problem. This caused quite a sensation, since there had been no serious attack on the problem for a great many years.

But it was not long before the experts began to cast aspersions on Hsiang's efforts. It seems that Hsiang had declared that "Thus and such is the worst possible configuration of balls and we shall content ourselves with examining this configuration." Unfortunately, Hsiang did not provide a convincing reduction to the indicated special case. In fact, it is generally believed that problems like this *do not have* a worst case scenario. One has to come up with arguments that address all cases at once. One of the leaders of the anti-Hsiang movement was Thomas Hales of the University of Michigan (now of the University of Pittsburgh). He published a polite but detailed article [HAL1] in the *Mathematical Intelligencer* taking Hsiang's efforts to task. It should be noted that others, including Conway, had endeavored to get Hsiang to admit to his errors. Conway published an important book [CON] on the subject and made no reference to Hsiang's work. But it was Hales who bit the bullet and made public statements to the effect that Hsiang was wrong. Hsiang replied to Hales's allegations in [HSI4].

Thomas Hales was not just blowing smoke. He had *his own* proof of the Kepler sphere-packing problem which he had intended to publish as a long sequence of papers. He studied the work of L. Fejes Tóth and concluded that the problem could be solved by minimizing a function of 150 variables. This in turn would entail solving about 100,000 linear

journals may take three years or more. But seven years is quite extraordinary. It must be borne in mind that over four years of this time was for the refereeing!

The current version of the truth is that the *Annals* has published an *outline* of Hales's proof.[97] Entitled *A Proof of the Kepler Conjecture*, this outline is 121 pages long (see [HAL2]). It is in the November, 2005 issue. The full details appeared elsewhere, in the *July, 2006* issue of *Discrete and Computational Geometry*. Hales's work took up the entire issue, and was divided into 10 papers, the final paper in 2010. These will comprise 265 pages.

This is also new territory for the estimable *Annals*. The *Annals of Mathematics* is, and has been, a stodgy old girl. She publishes complete, self-contained papers with complete proofs of new and important results. Its publication of Thomas Hales's work is charting new ground. And it is setting an example for other journals. It is possible that the entire nature of the publication of mathematical research will be affected by these actions.[98]

Those of a more technical bent are concerned about the Hales resolution of the Kepler problem because it uses `CPlex` to solve the relevant linear programming problems. The trouble with the software `CPlex` is that, while it is reliable, it does not certify any digit of any answer. Experts in numerical analysis consider this to be a design flaw. Perhaps this observation lends weight to the FlySpeck project (a 20-year program to computer-certify the Hales–Ferguson proof).

A fascinating and detailed history of the Kepler sphere-packing problem, and related mathematical questions, appears in [ASW]. Ferguson and Hales have been awarded the 2007 Robbins Prize by the American Mathematical Society, and that seems to be a fitting reward for their achievements.

10.4 Thurston's Geometrization Program

In 1866, Alfred Nobel (1833–1896) invented dynamite. This was in fact a dramatic event. It was just the technology that was needed for the Panama Canal and other big engineering projects of the day. Thus there was great demand for trinitrotoluene (TNT), and Nobel became a rich and successful man. He amassed quite a fortune. As is the case with many such wealthy and influential people, Nobel began, at the end of his life, to ponder his legacy. And he decided to create a prize to recognize the pinnacle of human achievement. It would entail considerable fanfare and honor for the recognized individual, and a substantial financial award. Thus was born the Nobel Prize. The prize was founded in Alfred Nobel's will in 1895, and it was first awarded in the areas of peace, literature, chemistry, physiology or medicine, and physics in 1901.

---

[97] This is also new territory for the venerable *Annals of Mathematics*. The *Annals* does *not* publish outlines.

[98] At a meeting, held in England, to discuss the changing nature of mathematical proof, MacPherson commented on the handling of the Hales paper by the *Annals*. He claimed that the *Annals* has a new policy of accepting computer-generated and computer-assisted proofs. There is no public record of such a decision. He further asserted that the usual refereeing process of the *Annals* had "broken down" in the case of the solution of the Kepler sphere-packing problem. The evidence, as described here, suggests that there was more at play than such a simple explanation would suggest.

Now Alfred Nobel was a practical man of the world. There was never any possibility that he would endow a Nobel Prize in metaphysical epistemology. It is likewise the case that he gave never a thought to mathematics. To this day, there is no Nobel Prize in mathematics. But history is a funny thing. For as long as anyone can remember, mathematicians have been telling each other that the reason that Nobel did not endow a prize in mathematics is that a mathematician ran off with Nobel's wife. This story may require some explanation.

A notable and celebrated contemporary of Alfred Nobel was Gösta Mittag-Leffler (1846–1927). Mittag-Leffler, a student of the celebrated mathematician Karl Weierstrass, was a prominent mathematician in his own right. He married well, and as a result lived in a grand mansion in Djursholm, Sweden—just outside of Stockholm.[99] Now Mittag-Leffler was a true celebrity; his name was in the newspapers all the time. He dressed like a dandy, and was really a man about town. Nobel was a dowdy, stodgy, solitary *bachelor*. He never married, and as far as we know he never had a lady friend in his entire adult life.[100] He was extremely jealous of Mittag-Leffler and the lifestyle that he led. It might also be noted that the extremely beautiful and brilliant mathematician Sonya Kovalevskaya was an associate of Mittag-Leffler and lived in his house for a period of time. It seems that their relationship was more than platonic. In any event, Mittag-Leffler represented a way of life that was an anathema to Nobel. Mittag-Leffler was the most prominent and celebrated scientist in all of Sweden. Some thought it likely that, were there a Nobel Prize in mathematics, Mittag-Leffler would have received it. This may have influenced Alfred Nobel's decision *not* to found a prize in mathematics.

In any event, there was and is no Nobel Prize in mathematics. Period.[101] This eventuality had an interesting upshot. Mittag-Leffler was quite irritated at the fact of no Nobel Prize in mathematics, since he too figured that he would be the obvious recipient of any such prize. So he used his considerable resources to establish the *Mittag-Leffler Prize* in Mathematics. And he specified explicitly that the medal for the prize would be twice as large as the medal for the Nobel Prize. Mittag-Leffler was the founder of the quite prestigious mathematical journal called *Acta Mathematica*. The winner of the Mittag-Leffler Prize would receive a complete leather-bound edition of the journal. Another aspect of the prize would be a lavish

---

[99] Djursholm is still, even to this day, a very ritzy place. Ordinary Swedes cannot afford to live there. In point of fact, the town is packed with the mansions of the ambassadors to Sweden from foreign countries.

[100] Some fairly recent revisionist biographies of Nobel depart from the version of his life that appeared in the "official" biography of himself that Nobel arranged and sanctioned. These newer books claim that, after Nobel's death, a woman came forward and filed a claim against Nobel's estate. She said that she had been, in effect, Nobel's common-law wife.

[101] It should be noted, however, that the Nobel Prize is an organic entity. It grows and changes. As an example, the Nobel Prize in Economics was added as recently as 1966. And even more recently than that, the Crafoord and Schock Prizes were added to the Nobel stable of awards. The Crafoord fund comes from a large pharmaceutical family, and the Schock money from a similar source. What is notable is that mathematicians are eligible for these two new prizes. Professor Louis Nirenberg of the Courant Institute of Mathematical Sciences was the recipient of the first Crafoord Prize. Professor Elias M. Stein of Princeton University has been awarded the Schock Prize. The money for these prizes is comparable in magnitude to that for the Nobel. Another new prize, which was first awarded in 2004, is the very prestigious Abel Prize. Sponsored by the Norwegian government, and named after the remarkable nineteenth-century Norwegian mathematician Niels Henrik Abel, this prize has emerged as the nearest thing to the Nobel Prize for mathematics.

banquet prepared by a famous French chef. Thus Mittag-Leffler strove to outclass the Nobel Prize in many different respects.

Unfortunately, whereas the Nobel Prize has survived and prospered for more than a century, the Mittag-Leffler Prize folded after being awarded only once or twice. The first time, it was awarded to Charles de la Vallée Poussin, a brilliant Belgian Fourier analyst. Professor de la Vallée Poussin was vacationing in the Alps at the time of the award, so it was arranged that the leather-bound copies of *Acta Mathematica* and the fancy French dinner be delivered to him up in the mountains.

As anybody who has thought about the matter would know, the way that a prize like this survives is that its founder invests the money. The awards are made each year from the income of the investment. That is where the Mittag-Leffler Prize met its sorry end. For Mittag-Leffler chose to invest in the Italian Railroad System and German World War I Bonds. End of the Mittag-Leffler Prize.[102]

One corollary of all this colorful history is that the mathematicians have created their own prize. Known as the Fields Medal, this prize was established by John Charles Fields of the Canadian Mathematical Society. Fields was in fact the President of the International Congress of Mathematicians that was held in Toronto in 1924, and was editor of the Proceedings. He suggested that the considerable funds raised through the sale of those Proceedings be used to establish a research prize for young mathematicians. In the 1932 meeting of the International Congress in Zurich, it was voted to approve this new award. It was dubbed the Fields Medal. The original stipulations for the award do not put many restrictions on who may be chosen for the prize. But it has become the custom that the Fields Medal should be awarded only to a mathematician under the age of 40: the recipient's fortieth birthday must not occur before January 1 of the year in which the Medal is awarded.

The purpose of the prize is to recognize talent in promising young mathematicians. First awarded in 1936—to Lars Ahlfors and Jesse Douglas—the prize began as a modest encomium to recognize developing talent. Over time *the Fields Medal has become the greatest honor that can be bestowed on a mathematician*. To win the Fields Medal is to be virtually beatified in mathematical circles. The set of *Fields Medalists* is a very select and distinguished group. Only 49 Fields Medals have been awarded in history—the most recent four in 2006.

The Canadian sculptor R. Tait McKenzie was enlisted to design the actual Fields Medal. The medal is 2.5 inches in diameter. The obverse side depicts the head of Archimedes facing to the right together with the Latin quotation "Transire suum pectus mundoque potiri" (To rise above oneself and grasp the world). The reverse side contains the Latin inscription

CONGREGATI
EX TOTO ORBE
MATHEMATICI
OB SCRIPTA INSIGNIA
TRIBUERE

which translates as "Mathematicians having assembled from the whole world, awarded for notable contributions."

---

[102] Well, not quite. A search of the Web reveals that there is still some vestige of the Mittag-Leffler Prize that is awarded these days. But it is nothing like the original prize.

William P. Thurston was awarded the Fields Medal in 1982 in Warsaw, Poland. He had done brilliant work on foliation theory (a branch of topology, which is part of the modern theory of geometry)—*in his Ph.D. thesis!!* These results completely revolutionized the field—solving many outstanding problems and opening up new doors. So the recognition was richly deserved. Thurston went on to do groundbreaking

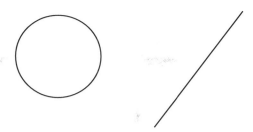

Figure 10.6. Classification of 1-manifolds.

work in all aspects of low-dimensional topology. He became a cultural icon for mathematicians young and old. He also had a great many brilliant students, and was able to spread his intellectual influence in that manner as well.

In 1977 (the first formal announcement was in 1982 in [THU1]), Thurston made a dramatic discovery. He had found a way to classify all 3-dimensional manifolds. Thurston's set of ideas was dubbed the "geometrization program." For a mathematician, a manifold is a surface with a specified dimension. This surface may or may not live in space; it could instead be an abstract construct. It was a classical fact from nineteenth century mathematics that all 1-manifolds and 2-manifolds had been completely understood and classified. The 1-manifolds are

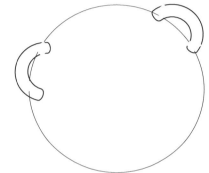

Figure 10.7. Classification of 2-manifolds

the circle and the line—see Figure 10.6. Any 1-dimensional "surface" is, after some stretching and bending, equivalent to a circle or a line. Any 2-dimensional surface (in space) is, after some stretching and bending, equivalent to a sphere with a certain number of handles attached; see Figure 10.7.

Thurston's daring idea was to break up any 3-dimensional manifold into pieces, each of which supports one of eight special, classical geometries.[103] Thurston worked out in considerable detail what these eight fundamental geometries must be. His theorem would give a structure theory for 3-dimensional manifolds.

Of course 3-dimensional manifolds are much more difficult to envision, much less to classify, than are 1- and 2-dimensional manifolds. Prior to Thurston, almost nothing was known about this problem. Three-dimensional manifolds are of interest from the point of view of cosmology and general relativity—just because we live in a 3-dimensional space. For pure mathematicians, the interest of the question, and perhaps the driving force behind

---

[103]In fact, it was the nineteenth-century mathematician Luigi Bianchi (1856–1928) who first identified these eight fundamental geometries. But Thurston saw further than Bianchi, or anyone else, insofar as to how they could be used.

the question, was the celebrated Poincaré conjecture. Formulated in 1904 by Henri Poincaré of Paris, it posited that any 3-dimensional surface with the geometry/topology of the sphere actually is equivalent to the sphere. Poincaré formulated this question following upon earlier investigations of homology spheres. He had some ideas about proving it, but they turned out to be inadequate to the task. The problem has fascinated mathematicians for 100 years. It has important implications for the geometry of our universe, and is of central interest to mathematicians and cosmologists alike. Every few years there is an announcement—that even makes it into the popular press—of a new proof of the Poincaré conjecture. In 1986, Colin Rourke of Warwick announced a proof. He is a man of considerable reputation, and his proof survived until it was dissected in a seminar in Berkeley. In 2002, M. J. Dunwoody of the University of Southampton announced a proof. He even wrote a 5-page preprint. The effort quickly died.

Many mathematicians have tried and failed to prove the Poincaré conjecture. But, if Thurston's geometrization program were correct, then the Poincaré conjecture would follow as an easy corollary. Suffice it to say that there was considerable excitement in the air pursuant to Thurston's announcement. He had already enjoyed considerable success with his earlier work—he was arguably the greatest geometer who had ever lived—and he was rarely wrong. People were confident that a new chapter of mathematics had been opened for all of us.

But the problem was with the proof. The geometrization program is not something that one proves in a page or two. It is an enormous enterprise that reinvents an entire subject. This is what historian of science Thomas Kuhn [KUH] would have called a "paradigm shift". Although Thurston was absolutely convinced that he could prove his new way of looking at low-dimensional geometry and topology (at least in certain key cases), he was having trouble communicating his proof to anyone else. There were so many new ideas, so many new constructs, so many unfamiliar artifacts, that it was nearly impossible to write down the argument. After a period of time, Thurston produced a set of "notes" [THU3] explaining the geometrization program.

It is important to understand what the word "notes" means to a mathematician. As this book explains and attests, mathematics has a long legacy—going back to Euclid and even earlier—of deriving ideas rigorously, using rigid rules of logic, and recording them according to the strict axiomatic method. Correctly recorded mathematics is crisp, precise, clear, and written according to a very standard and time-honored model. As a counterpoint to this Platonic role model, modern mathematics is a fast-moving subject, with many new ideas surfacing every week. There are exciting new concepts and techniques springing forth at an alarming rate. Frequently a mathematician may not wish to take the time to write down his/her ideas in a linear, rigorous fashion. If the idea is a big and important one, it could take a couple or several years to get the recorded version just right. Frequently one feels that he/she just doesn't have time for that. So a commonly used alternative is "notes." What the mathematician does is give a set of lectures, or perhaps a course (at the advanced graduate level) and get some of the students to take careful notes. The notes are rather quickly edited and then disseminated by the professor. We have already mentioned the Princeton math library's extensive collection of notes. Many a mathematician has cut his/her teeth, and laid the basis for a strong mathematics education by studying those notes.

And this is where William Thurston found himself around 1980.[104] He had one of the most profound and exciting new ideas to come along in decades. It would take him a very long time to whip all these new ideas into shape and shoehorn them into the usual mathematical formalism. So he gave some lectures and produced a set of notes. The Princeton University mathematics department, ever-supportive of its faculty, reproduced these notes and sold them to anyone—worldwide—who would send in a modest fee.

It is safe to say that these notes were a blockbuster. Many, many copies were sold and distributed all over the planet. There were so many beautiful new ideas here, and many a mathematician's research program was permanently affected or changed because of the new directions that these notes charted. But the rub was that nobody believed that these notes constituted a proof of the geometrization program. Thurston found this very frustrating. He continued to travel all over the world and to give lectures on his ideas, and to produce Ph.D. students who would carry the torch forth into the mathematical firmament. But he felt that this was all he could do—given the time constraints and limitations of traditional mathematical language—to get his ideas recorded and disseminated. The catch was that the mathematical community—which in the end is *always* the arbiter of what is correct and accepted—was not ready to validate this work.

In fact, Thurston's pique with this matter was *not* transitory. In 1994, he published the paper *On proof and progress in mathematics* [THU2], which is a remarkable polemic about the nature of mathematical proof. It also, *sotto voce*, castigates the mathematical community for being a bit slow on the uptake in embracing his ideas. This article was met with a broad spectrum of emotions ranging from astonishment to anger to frustration.

Many years later, Thurston published a more formal book [THU3]—in the prestigious Princeton University Press Mathematics Series—in which he began to systematically lay out the details of his geometrization program. In this tract he began at square one, and left no details to the imagination. He in effect invented a new way to look at geometry. His former Ph.D. student Silvio Levy played a decisive role in developing that book. And it is a remarkable and seminal contribution to the mathematical literature. In fact, the book recently won the important AMS Book Prize. But it should be stressed that this book is the first step in a long journey. If the full saga of Thurston's proof of the geometrization

---

[104]The Poincaré conjecture is one of those problems that frequently finds its way into the popular press. It is one of the really big problems in mathematics, and when someone claims to have solved it that is news. In the mid-1990s, Valentin Poénaru professed that he had a proof of the Poincaré conjecture. Poénaru is a professor at the University of Paris, and a man of considerable reputation. He produced a 1200-page manuscript containing his thoughts. Unfortunately, none of the experts were able to battle their way through this weighty tract, and no definitive decision was ever reached on Poénaru's work. In 1999 he published an expository tract with a summation of his efforts. Poénaru has contributed a number of important ideas to the subject of low-dimensional topology. But the jury is still out on whether he has proved the Poincaré conjecture.

In 2002, M. J. Dunwoody announced *his* proof of the Poincaré conjecture. The good news this time was that the paper that he wrote was only five pages long. Anyone could read it. But that was also the bad news. The fact that anyone could read it meant that it was not traditional, rigorous, hardcore mathematics, written in the usual argot. In fact, the paper was rather informal and fanciful. It was difficult to tell whether the work should be taken seriously (even though Dunwoody is a solid mathematician of good repute). In the end, Dunwoody's contribution was deemed flawed.

program is to be revealed in this form, then a great many more volumes must appear. And they have not appeared yet.

10.5 Grisha Perelman's Attack on the Poincaré Conjecture and the Geometrization Program

In 1982, mathematician Richard Hamilton introduced a new technique into the subject of geometric analysis. Called the method of "Ricci flows," this is a means of studying a flow generated from a source—like a heat flow. See Figure 10.8. What Hamilton does is to write down a differential equation on a given manifold that, in effect, mandates a notion of distance (what mathematicians call a "metric") and a motion of the manifold so that the velocity at any given point can be expressed in terms of the curvature. This idea is inspired perhaps by what happens to a closed, one-dimensional curve in the plane when you do the same thing to it: as the

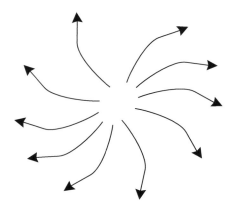

Figure 10.8. The Ricci flow.

process evolves, the curve smooths out, all its dents and twists and kinks disappear, and it becomes a circle. See Figure 10.9.

So this is what is also supposed to happen to a higher-dimensional manifold when it is subjected to the Ricci flow. The trouble is that things are much more complicated in higher dimensions. It is difficult to prove existence of a solution to the partial differential equation. And also some nasty singularities can arise during the deformation process (which we have somewhat whimsically depicted in Figure 10.9). Particularly tricky are the so-called "cigars" that are like long, thin tubes—see Figure 10.10. There is a sophisticated technique, developed at Princeton University by Bill Browder and John Milnor, called "surgery theory" that allows one to cut out these nasty singularities—as with a knife—and then to plug up the holes. The trouble was that, in the situation for the Poincaré conjecture, the singularities could run out of control. It was possible that one singularity could be removed and several others could spring up to take its place. (Grigori Perelman's deep insight, described in more

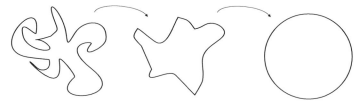

Figure 10.9. Singularities of the Ricci flow.

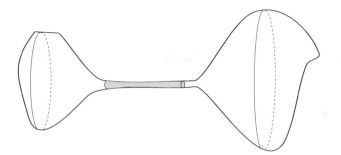

Figure 10.10.   A "cigar" in the Ricci flow.

detail below, was to be able to show that the singularities evolved in finite time. This gave the means for controlling them and, ultimately, eliminating them. The successive process of eliminating the singularities results in a nicer manifold—closer to the goal.)

Hamilton could see that this was potentially a method for solidifying the method initiated by Thurston in his "geometrization program". For one could start Ricci flows at different points of a surface and thereby generate the "geometric pieces" that Thurston predicted—refer to Figure 10.11. The idea of each of these special pieces is that it contains an "ideal geometry"—one in which a minute creature, equipped with a tape measure for the special metric on that piece, could not tell one point from another.

Hamilton was able to harness the analytic techniques to completely carry out this idea only for surfaces of dimension 2. (He made significant inroads in the case of dimension 3—enough to convince people that this was a potential program for proving the Poincaré conjecture—but some of the difficult estimates eluded him.)

Figure 10.11.   Ricci flow generates the geometrization decomposition.

And, as we have noted, 2-dimensional surfaces had already been classified by Jordan and Möbius in the mid-nineteenth century. Nothing new resulted from Hamilton's application of the Ricci flow to the study of 2-dimensional surfaces. He was able to obtain some partial results about existence of solutions for Ricci flows in dimension 3, but only over very small time intervals. He could make some interesting assertions about Ricci curvature, but these were insufficient to resolve the Poincaré conjecture in 3 dimensions. And this is where Ricci flows had stood for more than twenty years.[105]

---

[105] Hamilton has emerged as a hero in this story. He had the early idea that broke the Poincaré conjecture wide open. Although he could not himself bring the program to fruition, he was greatly admired both by Perelman and by Shing-Tung Yau. He gave the keynote address at the International Congress of Mathematicians in August 2006 about the status of the Poincaré conjecture.

Enter Grigori (Grisha) Perelman. Born in 1966, Perelman exhibited his mathematical genius early on. But he never had any particular designs to be a mathematician. His father, an engineer, gave him problems and things to read, but Perelman entered the profession crablike.

He won the International Mathematics Olympiad with a perfect score in 1982. Soon after completing his Ph.D. at St. Petersburg State University, he landed a position at the Steklov Institute in St. Petersburg (the Steklov Institutes are the most prestigious mathematics institutes in Russia). He accepted some fellowships at New York University and SUNY Stony Brook in 1992, and made such an impression that he garnered several significant job offers. He turned them all down. He impressed people early on that he was an unusual person. He kept to himself, let his fingernails grow to 6 inches in length ("If they grow, why wouldn't I let them grow?" said Perelman), kept a spartan diet of bread and cheese and milk, and maintained an eccentric profile.

In 1993 Perelman began a two-year fellowship at U. C. Berkeley. At this time he was invited to be a speaker at the 1994 International Congress in Zurich. He was also invited to apply for jobs at Stanford, Princeton, the Institute for Advanced Study, and the University of Tel Aviv. He would have none of it. When asked to submit his curriculum vitae for the job application, Perelman said, "If they know my work, they don't need to see my CV. If they need my CV, then they don't know my work."

In 1996 Perelman declined a prestigious prize for young mathematicians from the European Mathematical Society. It is said that Perelman claimed that the judging committee was not qualified to appreciate his work. But at this point everyone knew that Grisha Perelman was a man who was going places.

In the year 2002, on November 11 to be precise, Perelman wrote the groundbreaking paper *The entropy formula for the Ricci flow and its geometric applications* [PER1]. The fourth page of the introduction to this important paper contains the statement that in Section 13, he will provide a brief sketch of the proof of the (Thurston) geometrization conjecture. The "proof" was pretty sketchy indeed; and we are still unsure just what the paper [PER1] proves and does not prove. But the paper set the world aflame.

On November 19, 2002, geometer Vitali Kapovitch sent Perelman an email that read

> Hi Grisha, Sorry to bother you but a lot of people are asking me about your preprint "The entropy formula for the Ricci . . . " Do I understand it correctly that while you cannot yet do all the steps in the Hamilton program you can do enough so that using some collapsing results you can prove geometrization? Vitali.

Perelman's reply the very next day was, "That's correct. Grisha." Coming from a mathematician of the established ability and credentials of Perelman, this was a bombshell.

On March 10, 2003, Perelman released a second paper entitled *Ricci flow with surgery on three-manifolds* [PER2]. Among other things, this new paper filled in many of the details for the sketch provided in the first paper.

In April, 2003, Perelman gave a series of lectures at several high-level universities in the United States, including MIT (home of some of the most distinguished experts), SUNY at Stony Brook, New York University, and Columbia. These talks made a sufficiently strong impression that people began to take Perelman's program very seriously.

In July of that same year, Perelman released a third paper entitled *Finite extinction time for the solutions to the Ricci flow on certain three-manifolds* [PER3]. This paper provides a simplified version of a proof of a part of the geometrization program; that result is sufficient to imply the Poincaré conjecture.

One might cautiously compare Perelman's nine-month period—from November, 2002 to July, 2003—with Albert Einstein's "miracle year" (1905), in which he published four papers that completely changed the face of modern physics. In one of these papers Einstein introduced special relativity almost as an afterthought[106]—and it took several years for the idea really to catch on. But these were the four papers that really started everything. Just so for Perelman. His three papers profess to be able to harness the Ricci flow in dimension three. He claims to be able to prove the assertions of Thurston's geometrization program, and therefore also to be able to prove the Poincaré conjecture.

It must be stressed that Perelman's three papers are full of original and exciting ideas. But they are written in a rather informal style. And Perelman has no plans to publish them. They are posted on the Internet on the preprint server `arXiv` (see Section 8.3), and that is where they will remain. They will never be vetted or refereed, at least not in any formal sense. So it is up to the mathematical community to assess what these papers have to offer, and how they fit into the firmament. A delightful discussion, on an informal level, of the Poincaré conjecture and Perelman's contributions, appears in [STRZ].

One rarely sees in mathematics the level of excitement and intense activity that these three papers by Perelman have generated. There are conferences being organized all over the world. Such an august institution as the Clay Mathematics Institute in Cambridge, Massachusetts, funded two mathematicians (Bruce Kleiner and John Lott) at the University of Michigan to develop Perelman's ideas and produce a detailed and verifiable proof of the geometrization program. Two other very prominent mathematicians (John Morgan of Columbia and Gang Tian of Princeton, also with support from the Clay Institute) have given a large segment of their time and effort to writing up all the details of Perelman's program for the Poincaré conjecture; in fact their book [MOT] has now appeared.

There are a number of unusual threads to this story—some of which we have introduced in the last paragraph—that require explication. As we have discussed earlier, the Clay Mathematics Institute (refer to Section 9.2 for a discussion of mathematics institutes) has offered a $1 million dollar prize for the solution of each of seven seminal problems in modern mathematics. The Poincaré conjecture is one of these, and Perelman has in some sense proved it. He is the only living person who may have a claim on one of the Clay Prizes. So that is very exciting.

But there is the rub. For Perelman is not playing by any of the rules of the Clay Math Prizes. First of all, he has not published any of his three papers.

To make a long story short, Grisha Perelman put his seminal work on the Poincaré conjecture and the Thurston geometrization program on `arXiv` and nowhere else. He has no intention of publishing his papers in the traditional fashion. Thus there has been no refereeing or reviewing. And the papers, sad to say, are not written in the usual tight,

---

[106]This in the sense that relativity isn't even mentioned in the title!

rigorous, take-no-prisoners fashion that is the custom in mathematics. They are rather loose and informal, with occasional great leaps of faith.

Perelman himself has—in spite of receiving a number of attractive offers from major university math departments in the United States—returned to St. Petersburg so that he can care for his mother. He is more or less incommunicado, not answering most letters or emails. His view seems to be that he has made his contribution, he has displayed and disseminated his work, and that is all that he has to say. Perelman's position at Steklov pays less than $100 per month. But he lives an ascetic life. The jobs in the west that he declined pay quite prestigious (six-figure) salaries. Perelman claimed that he made enough money at his brief jobs in the west to support himself for the rest of his life.

But Perelman has left the rest of us holding the bag. The most recent information is that he has resigned his prestigious position at the Steklov Institute in St. Petersburg so that he can enjoy the solitude. He spends his time listening to opera and taking long walks. He no longer participates in mathematical life.

Perelman says that he is no longer a part of the mathematics profession. Because of a "competing" paper by Cao and Zhu (see below), and because of vigorous campaigning by various highly placed mathematicians, Perelman has concluded that the mathematics profession is deficient in its ethical code. He says that he has now quit mathematics to avoid making a fuss:

> As long as I was not conspicuous, I had a choice. Either to make some ugly thing or, if I didn't do this kind of thing, to be treated as a pet. Now, when I become a very conspicuous person, I cannot stay a pet and say nothing. That is why I had to quit.

All rather sad, and reminiscent of Fields Medalist Alexander Grothendieck (1928–). Grothendieck won the Fields Medal early on for his work (in functional analysis) on nuclear spaces. He later shifted interests, and worked intensely with his teacher Jean Dieudonné to develop the foundations of algebraic geometry. There is hardly any twentieth century mathematician who has received more honors, or more attention, than Grothendieck. He occupied a chair at the prestigious Institut des Hautes Études Scientifiques for many years—indeed, he was a founding member (along with Dieudonné). But, at the age of forty, Grothendieck decided to quit mathematics. Part of his concern was over government funding, but an equally large concern was lack of ethics in the profession. Even in 1988 Grothendieck turned down the prestigious Wolf Prize; his remarks at that time indicated that he was still disgusted with the lack of ethical standards among mathematicians. Today Grothendieck lives in the Pyrenees and is intensely introspective, to say the least. He believes that many humans are possessed by the devil.

In 2008 Perelman was awarded the Fields Medal at the International Congress of Mathematicians in Madrid (the other recipients were Andrei Okounkov, Terence Tao, and Wendelin Werner). This is without question the highest encomium in the profession.[107] Perelman did not show up to accept his honor; he formally (in advance) declined the award.

---

[107] Indeed, being selected for a Fields Medal is analogous to being canonized in the Catholic Church. And the selection procedure is just about as complicated.

In fact, the Fields Committee determined at the end of May 2006 to award Perelman (and three other mathematicians) the Fields Medal. President of the International Mathematical Union Sir John M. Ball traveled to St. Petersburg to endeavor to convince Perelman to accept the medal. Perelman was quite gracious, and spent a lot of time with Ball, but was adamant that he would not accept the award. He made it plain that what was important was solving the problem, not winning the prize.

Perelman does not want to publish his proof because then he would be a legitimate candidate for one of the seven Clay Millennium Prizes. And then, once he would have $1 millon, Perelman fears that some Russian gangster may then murder him. In 2010 he was awarded the Clay Millennium Prize, which he declined. He will not speak to the press and lives in seclusion.

At the International Congress of Mathematicians in 2008 Richard Hamilton gave the keynote plenary lecture. His purpose was to announce, and to reassure the mathematical community, that the Poincaré conjecture had indeed been proved and all was well. As part of his talk, Hamilton also announced that he himself had also an "alternative" proof of the Poincaré conjecture, from August, 2006. Late the following year in 2007 at another conference in China, Hamilton announced that he had written down about half of his proof, but he had run into problems. Nothing further was heard in this matter.

A footnote to the fascinating Perelman story is that just as this book was being written, Swiss mathematician Peter Mani-Levitska announced his own proof of the Poincaré conjecture. He wrote a self-contained, 20-page paper. And then he retired from his academic position and has not been heard from since. His proof uses combinatorial techniques that hark back to some of the earliest ideas (due to Poincaré himself) about the Poincaré conjecture. And Mani-Levitska, although not a topologist by trade, is one of the ranking experts in these combinatorial techniques. One version of the story is that select experts are now reviewing Mani-Levitska's proof. Another version is that the journal *Commentarii Mathematici Helvetici* has accepted the paper (suggesting that the paper has been refereed and verified by *someone*). The one sure fact is that most of the cognoscenti have been sworn to secrecy in the matter. It is difficult to get any hard information.

Another development is that Huai-Dong Cao and Xi-Ping Zhu published a 334-page paper that has appeared in the *Asian Journal of Mathematics*—see [CAZ]—which purported to prove *both* the geometrization program *and* the Poincaré conjecture. That is Fields Medalist S.T. Yau's journal, so this event carries some weight. It may be noted that the Cao/Zhu paper was published without the benefit of any refereeing or review. Yau obtained the approval of his board of editors, but *without* showing them the paper. On the other hand, Cao was a student of Yau. Yau is perhaps the premier expert in the application of methods of nonlinear partial differential equations in geometry. He presumably read the paper carefully, and that counts for a lot.

Yau aggressively promoted the Cao/Zhu work. Perelman himself has expressed some skepticism over what contribution this paper actually makes to the problem. He uses the paper as a touchstone to cast aspersions on the general ethical tenor in the mathematical profession. Unfortunately, it came to pass that the Cao/Zhu paper has been cast in a negative light. First, a portion of the paper was cribbed from the work of Kleiner and Lott. Some apologies have been tendered for that gaffe. The final resolution of that difficulty has yet to transpire.

Because of the very carefully written 473-page book of Morgan/Tian, it can now be said that Perelman's work has been carefully refereed and vetted. Two world-renowned experts have pronounced it (after some considerable ministrations of their own) to be correct. The entire Morgan/Tian book is available for purchase, and is also posted on the Web; it is therefore available for checking to the entire world. So it seems likely that the saga of the Poincaré conjecture has been brought to closure. Many experts say that we need to wait a while so that all the delicate, multidisciplinary aspects of the proof have time to gel and be fully understood. One wonders what Henri Poincaré himself might have thought of these developments that his problem has inspired, or what he would say about the ultimate solution.

The American Mathematical Society holds a large annual meeting, jointly with the Mathematical Association of America, each January. In 2007 that meeting was held in New Orleans, Louisiana. James Arthur, President of the AMS, had planned to have a celebration of the Poincaré conjecture at the gathering. A whole day of talks and discussions was planned. Fields Medalists John Milnor and William Thurston were to give background in the morning on the Poincaré conjecture and the geometrization problem respectively. In the afternoon, Richard Hamilton and John Morgan and John Lott were to speak about their contributions to the program. Unfortunately, Hamilton backed out; he cited other commitments and general fatigue. After considerable negotiation and cogitation, it was decided to invite Zhu to replace him. At that point Lott said he would not share the stage with Zhu. Efforts were made by AMS President James Arthur to rescue the situation, but to no avail. The event was ultimately canceled. There are hopes that a new celebration may be organized in the future.

11

A Legacy of Elusive Proofs

Modern mathematics is nearly characterized by the use of rigorous proofs. This practice, the result of literally thousands of years of refinement, has brought to mathematics a clarity and reliability unmatched by any other science. But it also makes mathematics slow and difficult: it is arguably the most disciplined of human intellectual activities. Groups and individuals within the mathematics community have from time to time tried being less compulsive about details of arguments. The results have been mixed, and they have occasionally been disastrous.

—Arthur Jaffe and Frank Quinn

There are no theorems in analysis—only proofs.

—John B. Garnett

Humble thyself, impotent reason!

—Blaise Pascal

Intuition is glorious, but the heaven of mathematics requires much more. . . . In theological terms, we are not saved by faith alone, but by faith and works.

—Saunders Mac Lane

And this prayer I make, Knowing that Nature never did betray The heart that loved her . . .
—William Wordsworth

11.1 The Riemann Hypothesis

Bernhard Riemann (1826–1866) was one of the true geniuses of nineteenth century mathematics. He lived only until the age of 39, ultimately defeated by poverty and ill health. He struggled all his life, and finally landed a regular professorship only when he was on his deathbed. But the legacy of profound mathematics that Riemann left continues to have a major influence in the subject.

When Riemann was taking his oral exams at Göttingen, Carl Friedrich Gauss assigned him to speak about geometry. And even the impatient and haughty Gauss was impressed; Riemann completely reinvented the subject. Taking into account the work of Bolyai

(1802–1860) and Lobachevsky (1793–1856) on non-Euclidean geometry, Riemann offered a visionary program for equipping surfaces and manifolds and even more general spaces with a geometry that retained the key features of the Euclidean geometry that we already know but also adapted itself to the particular features of the space in which it lived. Riemann's ideas about geometry are still studied intensively today.

Riemann's mathematical interests were broad. He gave the definition most commonly used today of the integral in calculus. He contributed fundamental ideas to the theory of trigonometric series and Fourier series. He made fundamental advances in the theory of complex variables. And he was definitely interested in number theory. In his seminal paper *On the number of primes less than a given magnitude* [RIE], Riemann laid out some key ideas about the distribution of the prime numbers. Recall that a prime number is a positive, whole number whose only whole-number divisors are 1 and itself. By custom we do not count 1 as a prime. The first several prime numbers are

$$2, 3, 5, 7, 11, 13, 17, 19, 23, 29, 31 \ldots .$$

The fundamental theorem of arithmetic tells us that every positive integer can be written in a unique manner as the product of primes. As an instance,

$$17640 = 2^3 \cdot 3^2 \cdot 5 \cdot 7^2.$$

Obviously the primes are the building blocks for everything we might want to know about the positive integers. The key ideas in modern cryptography are built on the primes. Many of the ideas in image compression and signal processing are founded in the primes. One of *the* most fundamental issues is how the primes are distributed.

When Gauss was a young man he studied tables of the prime numbers. These tables went on for pages and pages and contained thousands of prime numbers. From his studies, Gauss concluded that the following must be true: For n a positive integer, let $\pi(n)$ denote the number of primes less than or equal to n. For example, $\pi(50) = 15$, because the primes up to 50 are

$$2, 3, 5, 7, 11, 13, 17, 19, 23, 29, 31, 37, 41, 43, 47.$$

Also $\pi(100) = 25$ because, after 50, the primes are

$$53, 59, 61, 67, 71, 73, 79, 83, 89, 97.$$

Gauss conjectured that, for large n, the value of $\pi(n)$ is about $n/\log n$. More precisely, a refined version of Gauss's conjecture is that the limit of the quotient

$$\frac{\pi(n)}{n/\log n}$$

as n tends to infinity is 1.

Gauss was unable to prove this result, but he definitely believed it to be true. In fact, this so-called "prime number theorem" was not proved until 1896 (independently) by Jacques

Hadamard (1865–1963) and Charles de la Vallée Poussin (1866–1962). Their proof was remarkable because it used complex analysis (a subject that is rather distant from number theory both in form and in style) in a profound way. Central to their study of the prime number theorem was the notable zeta function of Bernhard Riemann.

In the aforementioned paper *On the number of primes less than a given magnitude,* Riemann introduced the zeta function. This is an analytic function of a complex variable that is defined by the infinite series

$$\zeta(z) = \sum_{n=1}^{\infty} \frac{1}{n^z}.$$

One of the first results Riemann proved about the zeta function was that it is intimately connected to the prime numbers. And Riemann *conjectured* that he knew the position of all the zeros (i.e., all the points of vanishing) of this zeta function. This was critical information for the study of the distribution of prime numbers. The celebrated Riemann hypothesis, perhaps the most important unsolved problem in modern mathematics, concerns the location of the zeros of the Riemann zeta function. The conjecture is that except for a collection of explicit and uninteresting zeros that were found at the negative even integers by Riemann, all the other zeros (and there are infinitely many of these) are located on the critical line—a vertical line in the Cartesian plane at $x = 1/2$. See Figure 11.1. G.H. Hardy (1877–1947) was able to show that infinitely many of these zeros are indeed on the critical line.

It is known that all the zeros of the zeta function—except for the trivial ones on the negative real axis that we noted above—lie in the critical strip, which is the set of complex numbers having real part between 0 and 1—refer to Figure 11.2. The issue is whether those zeros in the critical strip actually lie on the critical line. Brian Conrey of the American

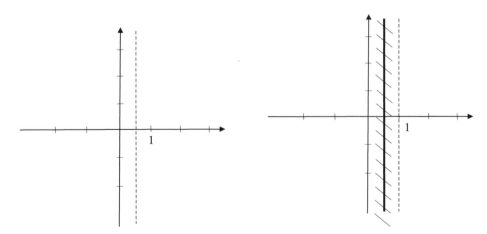

Figure 11.1. The critical line.   **Figure 11.2.** The critical strip.

████████████████████████████████

Riemann Hypothesis

The *Riemann hypothesis* is a conjecture about the location of the zeros of a certain analytic function of a complex variable. For a mathematician, that is a matter of considerable interest. But the Riemann hypothesis is particularly important because it implies important information about the distribution of the prime numbers. Today many techniques in cryptography (just as an instance) depend on such information. There are many important papers in all parts of mathematics that begin with the phrase "Assume that the Riemann hypothesis is true." or else with the phrase "Assume that the Riemann hypothesis is false." Whoever can establish the Riemann hypothesis will automatically be a great historical figure in our subject, and will also have a tremendous impact on all parts of mathematics.

████████████████████████████████

Institute of Mathematics has proved[108] that at least 2/5 of the zeros of the Riemann zeta function lie on the critical line.

The Riemann hypothesis has a colorful history (the book [SAB], for example, is a great source for detailed lore of the problem). Every major mathematician since the time of Riemann has spent time thinking about the Riemann hypothesis. In fact, the *illuminati* refer to the problem simply as RH. It is possible that—someday—a mathematician will imitate Andrew Wiles (see Section 10.8) and give a lecture after which he concludes by drawing a double arrow followed by "RH." That will be a dramatic day indeed.

One mathematician who had a real passion for RH was Godfrey Hardy. Every year he took a vacation to visit his friend Harald Bohr (1887–1951) in Denmark. The first thing they would do when they got together was to sit down and write an agenda. And, always, the first thing to go on the agenda was "Prove the Riemann hypothesis." This is one agenda item that they were never able to bring to fruition.

G.H. Hardy, a lifelong bachelor, had many eccentricities. He was convinced that he and God had an ongoing feud.[109] One corollary of this belief is that Hardy believed that God had an agenda against Hardy. When Hardy attended one of his beloved cricket matches he always brought an umbrella because he felt that this would guarantee that it would not rain: God would not give Hardy the satisfaction of being prepared for foul weather. One day Hardy had to make the passage across the English Channel—by boat, naturally—at a time of particularly rough seas. There was a genuine danger that he would not make it. As insurance, Hardy sent a postcard to Harald Bohr telling him that he had a proof of the Riemann hypothesis. This because Hardy was quite sure that God would not give him the satisfaction of going down with the ship and leaving the world to believe that he died with a proof of RH.

---

[108]Conrey built on earlier work that MIT mathematician Norman Levinson performed on his deathbed (Levinson in turn built on groundbreaking work by Fields Medalist Atle Selberg). Levinson at first thought that he could prove that 100% of zeros of the zeta function in the critical strip lie on the critical line. Then he had to modify his proof, and his claim, and make it 99%. Ultimately Levinson could prove that 33% of the zeros are on the critical line. Conrey's 2/5, or 40%, is a distinct improvement, and seems to take Levinson's method to its limit.

[109]The truth is that Hardy was an atheist. But he frequently joked about God.

Hans Rademacher (1892–1969) claimed in 1945 to have a disproof of the Riemann hypothesis. He was a distinguished mathematician at the University of Pennsylvania, and people took him very seriously. It is said that he explained the idea to none other than Paul Erdős, and Erdős gave it his blessing. In those days there was no Internet. And long-distance phone calls were considered to be prohibitively expensive. So Rademacher, as was the custom, prepared to send out postcards announcing his result. At the last minute Erdős urged him to be cautious, but the postcards went out anyway.

None other than Carl Ludwig Siegel found the flaw. The event did grace the pages of *Time* magazine, in the April 30, 1943 issue. The article contained a picture of Riemann with the caption, "Few understand it, none has proved it." The *Time* reporter wrote:

> A sure way for any mathematician to achieve immortal fame would be to prove or disprove the Riemann hypothesis . . . No layman has ever been able to understand it and no mathematician has ever proved it.
>
> One day last month electrifying news arrived at the University of Chicago office of Dr. Adrian Albert (1905–1972), editor of the *Transactions of the American Mathematical Society*. A wire from the society's secretary, University of Pennsylvania Professor John R. Kline, asked editor Albert to stop the presses; a paper disproving the Riemann Hypothesis was on the way. Its author: Professor Hans Adolf Rademacher, a refugee German mathematician now at Penn.
>
> On the heels of the telegram came a letter from Professor Rademacher himself reporting that his calculations had been checked and confirmed by famed mathematician Carl Ludwig Siegel of Princeton's Institute for Advanced Study. Editor Albert got ready to publish the historic paper in the May issue. US mathematicians, hearing the wildfire rumor, held their breath. Alas for drama, last week the issue went to press without the Rademacher article. At the last moment the professor wired meekly that it was all a mistake; on rechecking, mathematician Siegel had discovered a flaw (undisclosed) in the Rademacher reasoning. US mathematicians felt much like the morning after a phony armistice celebration. Said editor Albert, "The whole thing raised a lot of false hopes."

In modern times Louis de Branges (1932–   )—of Bieberbach conjecture fame (see Section 10.2)—was out to prove the Riemann hypothesis. Worse, Louis was up to his old tricks: He announced more than once that he had a proof. He even, more than once, produced a manuscript. But more than that is true. On one of these occasions he submitted an elaborate proposal to the National Science Foundation to sponsor a conference celebrating his proof of the Riemann hypothesis. All of de Branges's proofs have been found to be erroneous, and the conference was never funded.[110]

Unlike Fermat's last theorem, which Andrew Wiles worked on in secret because he feared that people would think him on a fool's errand, the Riemann hypothesis is definitely

---

[110]To be fair, there is some evidence that the troublesome parts of de Branges's argument are amenable to the knowledge and talents of certain mathematicians in Israel. The unfortunate political situation in that part of the world makes it infeasible for de Branges to take advantage of their talents (as he did in 1984 with the Soviet mathematicians in St. Petersburg). On the other hand, experts Brian Conrey and Xian-Jin Li have actually published a paper [COL] pointing out where the errors are in de Branges's arguments.

Now we proceed by crossing out all the multiples of 3 (but not 3 itself):

~~1~~	2	3	~~4~~	5	~~6~~	7	~~8~~	~~9~~	~~10~~	11	~~12~~	13	~~14~~	~~15~~	~~16~~
17	~~18~~	19	~~20~~	~~21~~	~~22~~	23	~~24~~	25	~~26~~	~~27~~	~~28~~	29	~~30~~	31	~~32~~
~~33~~	~~34~~	35	~~36~~	37	~~38~~	~~39~~	~~40~~	41	~~42~~	43	~~44~~	~~45~~	~~46~~	47	~~48~~
49	~~50~~	~~51~~	~~52~~	53	~~54~~	55	~~56~~	~~57~~	~~58~~	59	~~60~~	61	~~62~~	~~63~~	~~64~~
65	~~66~~	67	~~68~~	~~69~~	~~70~~	71	~~72~~	73	~~74~~	~~75~~	~~76~~	77	~~78~~	79	~~80~~
~~81~~	~~82~~	83	~~84~~	85	~~86~~	~~87~~	~~88~~	89	~~90~~	91	~~92~~	~~93~~	~~94~~	95	~~96~~

. . .

You can see that the numbers we are crossing out *cannot* be prime since in the first instance they are divisible by 2, and in the second instance, they are divisible by 3. Now we will cross out all the numbers that are divisible by 5 (why did we skip 4?) but not 5 itself. The result is

~~1~~	2	3	~~4~~	5	~~6~~	7	~~8~~	~~9~~	~~10~~	11	~~12~~	13	~~14~~	~~15~~	~~16~~
17	~~18~~	19	~~20~~	~~21~~	~~22~~	23	~~24~~	~~25~~	~~26~~	~~27~~	~~28~~	29	~~30~~	31	~~32~~
~~33~~	~~34~~	~~35~~	~~36~~	37	~~38~~	~~39~~	~~40~~	41	~~42~~	43	~~44~~	~~45~~	~~46~~	47	~~48~~
49	~~50~~	~~51~~	~~52~~	53	~~54~~	~~55~~	~~56~~	~~57~~	~~58~~	59	~~60~~	61	~~62~~	~~63~~	~~64~~
~~65~~	~~66~~	67	~~68~~	~~69~~	~~70~~	71	~~72~~	73	~~74~~	~~75~~	~~76~~	77	~~78~~	79	~~80~~
~~81~~	~~82~~	83	~~84~~	~~85~~	~~86~~	~~87~~	~~88~~	89	~~90~~	91	~~92~~	~~93~~	~~94~~	~~95~~	~~96~~

. . .

Let us perform this procedure just one more time, by crossing out all multiples of 7 (why can we safely skip 6?), but not 7 itself:

~~1~~	2	3	~~4~~	5	~~6~~	7	~~8~~	~~9~~	~~10~~	11	~~12~~	13	~~14~~	~~15~~	~~16~~
17	~~18~~	19	~~20~~	~~21~~	~~22~~	23	~~24~~	~~25~~	~~26~~	~~27~~	~~28~~	29	~~30~~	31	~~32~~
~~33~~	~~34~~	~~35~~	~~36~~	37	~~38~~	~~39~~	~~40~~	41	~~42~~	43	~~44~~	~~45~~	~~46~~	47	~~48~~
~~49~~	~~50~~	~~51~~	~~52~~	53	~~54~~	~~55~~	~~56~~	~~57~~	~~58~~	59	~~60~~	61	~~62~~	~~63~~	~~64~~
~~65~~	~~66~~	67	~~68~~	~~69~~	~~70~~	71	~~72~~	73	~~74~~	~~75~~	~~76~~	~~77~~	~~78~~	79	~~80~~
~~81~~	~~82~~	83	~~84~~	~~85~~	~~86~~	~~87~~	~~88~~	89	~~90~~	~~91~~	~~92~~	~~93~~	~~94~~	~~95~~	~~96~~

. . .

And now here is the punch line: The numbers that remain (i.e., that are *not* crossed out) are those that are *not* multiples of 2, or multiples of 3, or multiples of 5, or multiples of 7. In fact those that remain are not multiples of anything. They are the primes:

$$2, 3, 5, 7, 11, 13, 17, 19, 23, 29, 31, 37, 41, 43,$$
$$47, 53, 59, 61, 67, 71, 73, 79, 83, 89, \ldots$$

And on it goes. No prime was missed. The sieve of Eratosthenes will find them all.

But a number of interesting questions arise. We notice that our list contains a number of *prime pairs*: {3, 5}, {5, 7}, {11, 13}, {17, 19}, {29, 31}, {41, 43}, {71, 73}. These are primes in sequence that differ by just 2. How many such pairs are there? Could there be infinitely many prime pairs? To date, nobody knows the answer to this question. Another old problem is whether the list of primes contains arbitrarily long arithmetic sequences, that is, sequences that are evenly spaced. For example, 3, 5, 7 is a list of primes that is evenly spaced (by units of 2). Also 41, 47, 53, 59 is evenly spaced (by units of 6). It was only just proved in 2004 by Green and Tao that the primes *do contain* arbitrarily long arithmetic sequences.

An even more fundamental question is this: How many prime numbers are there altogether? Perhaps 100? Or 1000? Or 1,000,000? In fact Euclid determined that there are infinitely many prime numbers. We have discussed his argument elsewhere in this book (Subsection 2.2.1).

So this is what a sieve technique is: a systematic method for crossing off elements of the positive integers in order to leave behind the numbers that we seek. Many of the known attempts at the Goldbach conjecture have involved some sort of sieve technique. Although this problem is not as central or important as the Riemann hypothesis, it has attracted considerable attention. No less a vanguard than Brun, Rademacher, Vinogradov, Chen, Hua, Bombieri, and Iwaniec have worked on Goldbach and also on the twin-prime conjecture (discussed in the next section). Perhaps what is most important about these problems is not the problems themselves, but rather the techniques that have been introduced to study them.

11.3 The Twin-Prime Conjecture

The twin-prime conjecture asks another tantalizing question that is accessible to anyone. We call two prime numbers "twin primes" if there is a difference of just 2 between them. Examples of twin primes are

$$\{3, 5\}, \{5, 7\}, \{11, 13\}, \{17, 19\}, \{29, 31\}, \{41, 43\} \ldots.$$

The twin primes become sparser and sparser as we move out into the larger integers. The question is: are there infinitely many twin primes?

Again, computers can be used to find twin primes that are quite far out in the number system. Billions of twin primes have been found. And the problem can be attacked using sieve methods; it has been subjected to such assaults on many occasions. It may be noted that Vinogradov has pioneered the method of trigonometric sums (based on ideas of Hardy and Littlewood) in the study of the twin-primes conjecture.

The number theorist Viggo Brun proved in 1919 the startling result that the sum of the reciprocals of all the twin primes is a convergent series. This is particularly striking because it is a classical result (proved by Euler in the eighteenth century) that the sum of the reciprocals of *all* the primes diverges. The latter gives a new (and very different proof from Euclid's) that there are infinitely many prime numbers.

In 1966, Jingrun Chen showed that there are infinitely many primes p such that $p + 2$ is either a prime or a semiprime (i.e., the product of two primes). The approach he took involved sieve theory, and he managed to treat the twin prime conjecture and Goldbach's conjecture in similar manners.

The trouble with both the Goldbach conjecture and the twin-prime conjecture is that neither is very central to modern mathematics. They are more like historical footnotes. Still, it must be noted—once again—that a number of eminent mathematicians have worked on these problems. It is difficult to judge their true worth until we know that they are true and can spend some time with them.

Fermat's last theorem loomed large—in an obvious sort of way—because it spawned so much important mathematics. Ring theory, ideal theory, and many other critical ideas of modern abstract algebra grew out of attempts to prove Fermat's last theorem. The twin-prime conjecture has not been nearly so fecund, although it definitely has been the wellspring of some interesting research and some new ideas.

The mathematician who solves either of these old chestnuts (Goldbach or twin-prime) will achieve a notable amount of celebrity, and will reap particular rewards. And it is fitting that this will be so. But it will not be like proving the Riemann hypothesis or the Poincaré conjecture.

11.4   Stephen Wolfram and *A New Kind of Science*

Stephen Wolfram is one of the *Wunderkinder* of modern science. He earned his Ph.D. at Caltech at the age of 20. Just 1 year later he was awarded a MacArthur Fellowship—the youngest recipient ever! Most people who are granted a MacArthur Fellowship—which is quite substantial (on the order of $500,000 or more)—just stick the money in the bank and stare at it. But Wolfram is quite the enterprising fellow, and he started a company (now called Wolfram Research) and developed one of the first computer algebra systems for the personal computer (see Sections 7.2 and 8.2 for further discussion). Known as Mathematica, this product has really changed the landscape for mathematical scientists. Whereas formerly one used a computer to do strictly *numerical* calculations, now the computer can solve algebra problems, solve differential equations, calculate integrals, calculate derivatives, manipulate matrices, and perform many other basic mathematical operations. Mathematica can draw marvelous graphs and diagrams. Today there are many thousands of mathematicians who depend on Mathematica as a basic tool in their research. And, concomitantly, Stephen Wolfram is a wealthy man.

As already noted, Wolfram is an estimable scientist, and a man with diverse talents. When he was on the faculty at the University of Illinois, he was simultaneously a professor of computer science, mathematics, and physics—a feat hardly ever equaled in the annals of academe. One of his major contributions to mathematical science has been to give a prominent berth to the theory of cellular automata. Just what is this animal? A cellular automaton is a graphic system that begins with an array of boxes on a sheet of graph paper and a set of rules by which these boxes might evolve. For example, one might begin with a single box as shown in Figure 11.3. Note that the key box is shaded, and the other boxes shown are *un*shaded. Then the rule might be as depicted in Figure 11.4.[111] The way to read this rule is as follows. The first box shows a particular triple of boxes: they are all shaded.

---

[111] This is in fact Wolfram's Rule 90—see [WOL].

Whenever we encounter such a triple then the middle box is to be made blank. The second box shows a new triple in which the first two boxes are shaded and the third is not. Whenever we encounter such a triple then the middle box is to be shaded. And so forth. Note that when we are applying this analysis to the Nth row on the graph paper, we record the result in the $(N + 1)$st row.

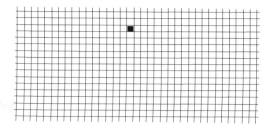

Figure 11.3. Beginning of a cellular automaton.

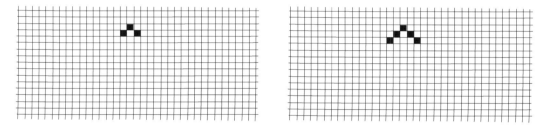

Figure 11.4. Wolfram's cellular automaton rule 90.

Figure 11.5. First iteration of rule 90.

Figure 11.6. Second iteration of rule 90.

Applying this rule to our initial configuration in Figure 11.3 (parts 7, 6, and 4 of the rule), we obtain Figure 11.5. Now apply the rule again to obtain Figure 11.6. And yet once more to obtain Figure 11.7. And so forth.

The remarkable thing about a cellular automaton is that one can begin with a very simple configuration and a very simple rule of generation and end up, after a good many iterations of the rule, with something fantastically complicated. Figures 11.8, 11.9, and 11.10 exhibit some configurations that were generated by fairly simple cellular automata.[112]

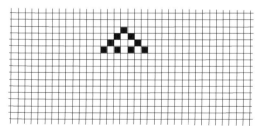

Figure 11.7. Third iteration of rule 90.

[112]John Horton Conway's game of life has popularized cellular automata, and put them in the pubic eye.

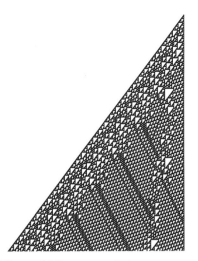

Figure 11.8.   First cellular automaton.

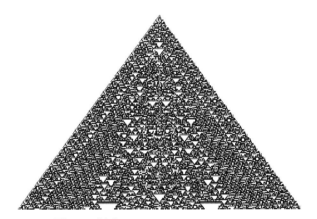

Figure 11.9.   Second cellular automaton.

Over time, Wolfram has become obsessed with cellular automata. He is convinced that they are a model for how nature works. From the spots on a leopard to the design of a snowflake to the structure of the human brain, Wolfram believes that there is a cellular automaton that encodes the design of each. He has written a 1286-page book [WOL] that explains his theory in some detail. The book has garnered quite a lot of attention—see the review [KRA6] as well as the Web site

```
http://www.math.usf.edu/~eclark/ANKOS_reviews.html
```

But *A New Kind of Science* is *not* Newton's *Principia* and it is not Darwin's *On the Origin of Species*. Whatever its merits may be, this book has not changed anyone's view of

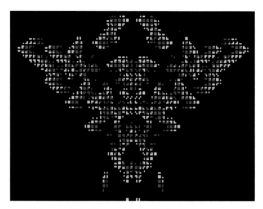

Figure 11.10.   Third cellular automaton.

the world. It has not changed any school curriculum (although Wolfram has written that it will). It has not convinced anyone of anything.

Given the message of the present book, and given that the theory of cellular automata is mathematical science, the subject merits some discussion in these pages. Wolfram is a physicist by training. He is not bound to the strict rigor of proofs as we have discussed it here. But he knows a heck of a lot of mathematics. He knows what the standards are; and he is well acquainted with the standards in the world of theoretical physics. Unfortunately, the book [WOL] does not meet any of these standards. How could this be, and why would Wolfram commit such a gaffe?

What Wolfram professes in the pages of [WOL] is that he is writing for a popular audience. He thinks that his ideas are so important that he must leapfrog over the usual jury of his scientific peers and go directly to the populace at large. Unfortunately, one result of this decision is that he cannot write with any rigor or sophistication. He cannot indulge in any precise proof or argumentation. He cannot access and utilize the scientific literature. The bottom line is then a panorama of rather vague discussion that is, to be generous, quite inconclusive. The considerations in [WOL] are largely phenomenological, and they are *not* convincing.

Wolfram's book contains a *great many* calculations, a *great many* pictures, and a *great many descriptions*. But it has very little *scientific discourse*. Because Wolfram wants to bypass the usual jury of scientific reviewers, he must express himself in a language that is comprehensible to virtually anyone. And that imposes severe limitations on what he can do and what he can say. He cannot assume that his readers know the basic tenets of physics; he cannot assume that they know relativity, or quantum mechanics, or string theory. He cannot even assume that they know calculus (which is, after all, the bedrock of modern physics). The upshot is that he can only dance around the key ideas. He cannot truly address any of them.

And this is a very important point. Scientific discourse is hard-edged, it is difficult, it is unforgiving. Rank amateurs simply cannot read the scientific literature. They do not have the training and they do not have the know-how. But this is not a condemnation of science, and it is not a condemnation of scientific discourse. It is just a fact that science is hard and technical. It requires some sophistication to understand the *lingua franca*. By trying to leap over this divide, Wolfram has found himself in no-man's land. The scientists cannot understand what he is talking about because he too vague and imprecise. The laymen cannot understand what he is talking about because the subject matter simply does not lend itself to casual, imprecise description. He has chosen the wrong medium for his message. It was preordained that it could not work.

And that is the lesson for the readers of this book. Writing and reading strict analytical proofs, or any kind of rigorous deduction, is hard work. But, in the end, the hard work pays off. A formal proof is comprehensible to anyone with the proper training. And it reads accurately in all countries and all cultures at all times. It will be just as valid, and just as comprehensible, in 100 years as it is today. Wolfram's ideas garnered some short-term attention. But they are already fading into the sands of time.

11.5   Benoît Mandelbrot and Fractals

One of the notable phenomena of modern mathematics is the development of fractal theory. Spawned by Benoît Mandelbrot (1924–2010) in 1982, fractal geometry is touted to be a new blueprint for understanding the world around us. Some pictures of fractals are shown in Figures 11.11, 11.12, and 11.13. In fact, Figure 11.13 exhibits the famous Mandelbrot set.

Mandelbrot begins his famous book [MAN1] by pointing out that

> Clouds are not spheres, mountains are not cones, coastlines are not circles, and bark is not smooth, nor does lightning travel in a straight line.

Instead, he posits, the objects that occur in nature are very complicated. Typically they have very wiggly boundaries. The geometric phenomenon on which he focuses is an object such that if you take a piece of it and blow it up (i.e., dilate it), it looks the same. See Figure 11.14.

Figure 11.11.   First fractal.

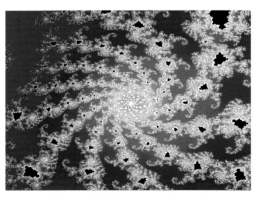

Figure 11.12.   Second fractal.

Fractals

A *fractal* is a geometric object that enjoys certain self-similarity properties. If you magnify a part of the fractal then you get an isomorphic image of the original object. Benoît Mandelbrot is usually credited with the invention of fractals.

Many fractals have a fractional dimension, and that fact adds to their mystique. Thanks to mathematician John Hubbard, we now have wonderful ways to render color images of various fractals on the computer.

Fractals have been used to model a variety of notions in physics and engineering. They are used daily to help detect faults and fissures in structures, and they are used in visual imaging. They have really changed our view of mathematics and of the physical world.

Fractal geometry is now a huge industry, and there are those who use fractals to create denoising filters, implement image-compression algorithms, and analyze investment strategies.

Mandelbrot claims not only to have invented a new branch of mathematics (in fact he did not actually invent the Mandelbrot set—it appears earlier in [BRM]). He also asserts that he has created a new way of *doing* mathematics. He has been known to deride traditional mathematicians who do mathematics in the traditional fashion (see [MAN2]). Mandelbrot's new methodology, reminiscent of Wolfram's study of cellular automata, is phenomenological. The fractal geometer typically does not enunciate theorems and prove them. Instead the fractal geometer generates computer graphics and describes them and endeavors to draw conclusions.

One can easily imagine that fractal geometry is extremely popular among mathematical amateurs and also among mathematically knowledgeable teachers. Amateurs like fractals because they feel that the subject gives them a glimpse of some current mathematics—the grist of cutting-edge research. High school teachers

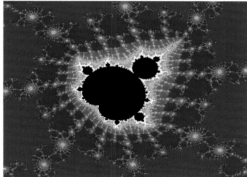

Figure 11.13. The Mandelbrot set

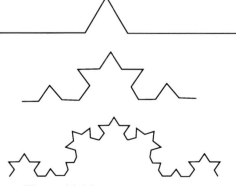

Figure 11.14. Scalability of fractals.

like fractals because they feel that they can show their students some genuine, modern mathematics—just by showing them pictures of fractals and the discussion adhering thereto. For a number of years Mandelbrot (an employee of IBM) was the face of research of International Business Machines. IBM's primetime television ads proudly featured Mandelbrot and pictures of fractals. And IBM rewarded Mandelbrot accordingly. The company used its financial clout to get Mandelbrot a teaching position at Yale University and the prestigious Barnard Prize, among other encomia.

It must be said that there are many branches of mathematics and science in which fractals are considered to be extremely useful. Physicists are always looking for new ways to model nature, and fractals give them a new language for formulating the laws of our world. More than half the submissions to some of the major physics journals concern fractals. Statisticians in Australia use fractals to model semipermeable membranes. Mandelbrot himself used fractal ideas to help AT&T with a denoising problem.

Fractal geometry is both a new way to conceive of shape and form and also a new way to think about mathematics. It does *not* conform to the Euclidean paradigm of theorem and proof. It in fact epitomizes a new experimental approach to mathematics that is complementary to the more familiar and traditional methodologies.

11.6 Roger Penrose and *The Emperor's New Mind*

In the mid-1980s, distinguished British physicist Roger Penrose of Oxford University was watching the "telly"—absorbing a show about artificial intelligence. The show seemed to be claiming that soon machines will be able to think just as human beings do. These assertions pushed all of Roger Penrose's buttons—all the *wrong* buttons. He went marginally apoplectic, set aside all his research projects, and decided that he had to write a book refuting these claims. The result was *The Emperor's New Mind* [PEN1], a book that has had a decisive impact on modern thinking. It was a bestseller to boot, in spite of the fact that it is so profoundly technical that there are few university professors who can read it straight through.

Penrose is a remarkable scholar. He has advanced training in mathematics, and he has made significant contributions to modern mathematical thought. But his primary work for the past several decades has been in theoretical physics. He is a close associate of Stephen Hawking. Penrose's expertise spans several academic disciplines. And his book reveals his erudition in considerable detail.

To oversimplify his thesis, Penrose says in his book that a computer can only execute a procedure that is given by an algorithm—typically a *computer program*. But a mathematician, for example, finds mathematical truths by way of insight, intuition, and sometimes even leaps of faith. These mental operations *cannot* be achieved or implemented by means of an algorithm. Penrose might have gone further to note that the creation of artwork—whether it be poetry, music, literature, sculpture, dance, or painting—is not generally achieved by way of an algorithm. It is generally done heuristically, with a basis in the artist's life experience.

The intellectual, scholarly embodiment of the thesis that machines can, or some day will be able to, think is the subject of "artificial intelligence," or AI. We have enjoyed

AI at our universities for only forty years or so—thanks largely to vigorous support from the military establishment. It is easy to see why military leaders, those who think about armaments and warfare, would like to have a machine that can look down the road and tell what is coming. They would like to have a machine that can evaluate a battle situation and determine how to deploy troops or armaments. They would like to have a machine that can produce optimal bombing strategies. For a good many years (stretching back to the days of Sir Walter Raleigh), the aiming and disposition of artillery has been one of the prime motivators for various mathematical calculations, for the construction of a variety of computing machines, and for the development of certain branches of mathematics (such as spherical trigonometry and geodesy). Now we are in a much more sophisticated age, and the demands of technology are considerably more recondite.

It can be argued, and it is frequently proposed by the skeptics and the nay-sayers, that AI has achieved little. To be sure, we now have some rudimentary robots. We have a machine that can walk through the room without bumping into the furniture. We have a "robot" that can vacuum your living room or mow your lawn. Much modern manufacturing—the sort of rote work that used to be performed by assembly-line workers—is now done by robots. But there is no robot that can tie a shoe.

The field of AI has spawned some other fields that play a notable role in modern theoretical computer science; "expert systems" is one of these. But there are no machines today, nor will there be any in the foreseeable future, that can *think*.

The discourse in which Roger Penrose engages is quite tricky. For, reasoning like a mathematician, one must decide in advance what it means to "think." Questions about Gödel's incompleteness theorem and Turing machines are natural to consider. Church's thesis, a venerable tenet of modern logic, is that any effectively computable function is a recursive function. What does this mean? A *recursive function*—this is an important idea of Kurt Gödel—is a function that is built up from very basic steps—steps that a machine can perform. There is a technical, precise, mathematical definition of recursive function that can be found in textbooks—see [WOLF] as well as [KRA4] and references therein. It is rather more difficult to say what an effectively computable function is—much of the modern discourse on Church's thesis concerns trying to come up with the right definition of this concept. Roughly speaking, an *effectively computable function* is one that a machine can calculate. The intuition behind Church's thesis is that any "effectively computable function" must be one that can be broken down into simple, algorithmic steps. Hence it must be recursive.

One approach to the questions being considered here is that any effectively computable function, or any procedure prescribed by a Turing machine, is an instance of human thought. And it is clear by their definitions that these are in fact operations that a computer could perform. But Penrose would argue that this is an extremely limited perception of what human thought actually is. Beethoven's *Ninth Symphony*, *War and Peace*, even the invention of the computer itself, would never have been achieved by way of such mechanized or procedural thinking. Creative thought does not proceed from algorithms.

It is also natural to consider Penrose's concerns from the point of view of the Gödel incompleteness theorem. In any logical system that is at least as complex as arithmetic, there will be true statements that we cannot prove (inside the system). Thus a computer will always be limited, no matter what language or logical system it is using, in what it can achieve. A human being can deal with the "Gödel sentence" heuristically, or by amassing evidence, or by offering a plausibility or perhaps a probabilistic argument. The human being can step outside the logical system and exploit whatever tools may be needed to obtain a proof. The computer cannot deal with the problem at all. Computer scientists (see [MCC]) like to point out that the programming language `Lisp` is very good at handling the Gödel phenomenon. And we can also construct systems that will simply avoid Gödel sentences. Nevertheless, Gödel's incompleteness argument carries genuine philosophical weight.

It is easy to imagine that many classically trained theoretical mathematicians are more than anxious to embrace Penrose's ideas. There are few of us who want to believe that we shall some day be supplanted by a room full of machines running on silicon chips. But an equally enthusiastic cadre from the computer science community are quick to oppose Penrose. The rather articulate and entertaining article [MCC] is an instance of the kind of arguments that have been aligned against Penrose.

The AI community continues to flourish—MIT and Caltech, for example, have vigorous artificial intelligence groups. There are obviously many interesting questions to consider in this discipline. But it is noteworthy to attend a gathering of AI investigators and to see how much of their time they spend arguing over definitions. As mathematicians, we appreciate the importance of laying down the right definitions so that everything else will follow rigorously therefrom. But we allot only a finite amount of time to that effort and then we move on. The subject of artificial intelligence has a more organic component to it—after all, it is an effort to apply mathematical principles to the way that the human mind functions—and therefore it is subject to forces and pressures from which theoretical mathematics is immune. Roger Penrose touched a nerve with his thinking about the efforts of the AI community. His ideas continue to reverberate throughout the mathematical sciences.

11.7 The P/NP Problem

Complexity theory is a device for measuring how complicated a problem is (from a computational point of view), or how much computer time it will take to solve it. As this book has made clear, computation (and the theory of computation) looms large in modern mathematical science. Of particular interest is the question of which problems can be solved in a reasonable amount of time, and which will require an inordinate or impractical amount of time. Problems of the latter sort are termed *computationally expensive*. Problems of the former type are called *feasible*. Just as an instance, the problem of factoring a whole number with 200 digits is computationally expensive. This could actually take many years, even on a fast digital computer. But if instead I say to you:

> Here is a 200-digit number N. I claim that these two 100-digit numbers p and q are its prime factors. Please verify this assertion.

Then all you need to do is to multiply p and q together. This will just take a few seconds (in fact typing in the numbers is the part of the problem that takes longest).

The issues being considered here are part of the **NP**-completeness problem. Many people consider this to be the most important problem in mathematics and theoretical computer science. It has far-reaching consequences for what we can do with computers, or what we can hope to do in the future. Many of the fundamental ideas in cryptography depend on questions of computational complexity. We shall discuss some of these ideas in the present section.

11.7.1 The Complexity of a Problem

Complexity theory is a means of measuring how complicated it is, or how much computer time it will take, to solve a problem. We measure complexity in the following way: Suppose that the formulation of an instance of a problem involves n pieces of data. Then how many steps[113] will it take (as a function of n) to solve the problem? Can we obtain an effective bound on that number of steps that is valid for asymptotically large values of n?

Consider dropping n playing cards on the floor. Your job is to put them back in order. How many steps will this take?

In at most n steps (just by examining each card), you can locate the first card. In at most another $(n-1)$ steps, you can locate the second card. In at most another $(n-2)$ steps, you can locate the third card, and so forth. In summary, it will require at most

$$n + (n-1) + (n-2) + \cdots + 1 = \frac{n(n+1)}{2}$$

steps to put the deck of cards back in order (this summation formula is usually attributed to Gauss). Because the expression $n[n+1]/2 \leq n^2$, we say that this problem has *polynomial complexity of degree (at most) 2*—where 2 is the exponent of the majorizing monomial.

Now contrast that first example with the celebrated traveling salesman problem: There are given n cities and there is a road connecting each pair (see Figure 11.15), with a cost assigned to each of these roads. The problem is to find the route that will enable the salesman to visit each city, to return to his starting place, and at the least aggregate cost.

On the level of effective computability we see that a search for the solution of the traveling salesman problem amounts to examining each possible ordering of cities (because clearly the salesman can visit the cities in any order). There are $n! = n \cdot (n-1) \cdot (n-2) \cdots 3 \cdot 2 \cdot 1$ such orderings. According to Stirling's formula,

$$n! \approx \left(\frac{n}{e}\right)^n \cdot \sqrt{2\pi n}.$$

Thus the problem cannot be solved (in the obvious way) in a polynomial number of steps. We say that the problem (potentially) has *exponential complexity*. It actually requires some additional arguments to establish this fact.

Another famous problem that is exponentially complex is the *subgraph problem*. Let G be a graph; that is, G is a collection of vertices together with certain edges that connect certain pairs of the vertices. Let H be another graph. The question is whether G contains

---

[113]Here a "step" is an elementary action that a computer or robot could take. It should be a "no brainer".

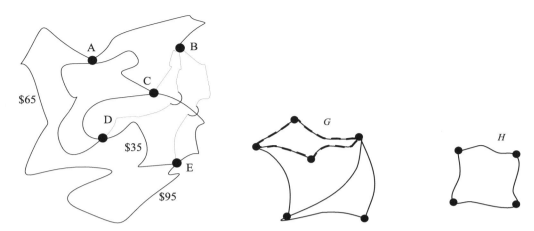

Figure 11.15.   The traveling salesman problem.

Figure 11.16.   Subgraph problem.

a subgraph that is isomorphic to H (Figure 11.16). (Here two graphs are isomorphic if there is a combinatorial mapping matching up vertices and edges.)

11.7.2   Comparing Polynomial and Exponential Complexity

From the point of view of theoretical computer science, it is a matter of considerable interest to know whether a problem is of polynomial or of exponential complexity, for this information gives an indication of how computationally expensive a certain procedure will be. In the area of computer games (for example), this will translate into whether a certain action can be executed with realistic speed, or sluggishly.

In the discussion that follows, we restrict attention to the so-called "decision problems." These are problems with yes/no answers. An example of a problem that is *not* a decision problem is an optimization problem (e.g., find the configuration that maximizes something). But in fact most optimization problems can be converted to decision problems by introducing an auxiliary parameter that plays the role of an upper bound. It is not a severe restriction to treat only decision problems, and it makes the exposition much cleaner.

11.7.3   Polynomial Complexity

A problem is said to be of "Class **P**" if there are a polynomial p and a (deterministic) Turing machine (DTM) for which every input of length n comes to a halt, with a yes/no answer, after at most $p(n)$ steps (see Section 7.1 for a discussion of Turing machines). The word "deterministic" is used here to denote an effectively computable process, with no guessing (see Subsection 11.7.5 below for further consideration of these ideas).

Problems of Class **P** are considered to be tractable. A problem that is not of class **P**—that is, for which there is no polynomial solution algorithm—is by definition *intractable*. Class **P** problems give solutions in a reasonable amount of time, and *they always give* *solutions*. It will never happen that the machine runs forever. The problem of ordering a deck of cards (already discussed), the problem of finding one particular marble in a jar of marbles, and the problem of matching up husbands and wives at a party are all problems of Class **P**.

11.7.4 Assertions That Can Be Verified in Polynomial Time

More significant for the theoretical development of this section is that, for certain problems that are otherwise intractable, the *verification of a solution* can be a problem of Class **P**. We describe the details of this assertion below.

For example, the problem of finding the prime factorization of a given natural number N with n digits is believed to be of exponential complexity (see [SCL]). But in fact this is not known, as of this writing. More precisely, the complexity—according to the best currently known algorithm—is about of size $10^{\sqrt{n \ln n}}$, which means that the computation would take about that many steps. But the verification procedure is of polynomial complexity: If N is given and a putative factorization $p_1, \ldots, p_k$ is given, then it is obviously (just by inspection of the rules of arithmetic) a polynomial-time problem to calculate $p_1 \cdot p_2 \cdots p_k$ and verify (or not) that it equals N.

The problem of factoring a large integer is believed by all the experts to be of exponential complexity (that is, if the integer has n digits then it will take on the order of 2^n steps to find the prime factorization).[114] The celebrated RSA encryption method—one of the cutting-edge techniques in modern cryptography—is premised on this hypothesis. That is, if we could find a way to factor large integers in polynomial time, then RSA-encrypted messages could be rapidly decrypted.[115] At this time it is not known whether the factorization of large integers is a polynomial-time problem. The best algorithms that we have are exponential time algorithms.

Likewise, the subgraph problem is known to be of exponential complexity. But, if one is given a graph G, a subgraph K, and another graph H, then it is a problem of only polynomial complexity to confirm (or not) that H is graph-theoretically isomorphic to K.

The considerations in the last three paragraphs will play a decisive role in our development of the concept of "problems of class **NP**."

11.7.5 Nondeterministic Turing Machines

A nondeterministic Turing machine (NDTM) is a Turing machine with an extra write-only head that generates an initial guess for the Turing machine to evaluate (see [GAJ] for a rigorous definition of the concept of nondeterministic Turing machine). We say that a problem is of Class **NP** if there are a polynomial p and a nondeterministic Turing machine with the property that for any instance of the problem, the Turing machine will generate a

---

[114]It is an amazing recent result of Agrawal and his group [AKS] that there is a polynomial-time algorithm for determining whether a given positive integer is prime or composite. But Agrawal's method will *not* produce the prime factorization. It only provides a "yes" or "no" answer.

[115]The amusing film *Sneakers*, starring Robert Redford, is in fact precisely about such an eventuality.

Fermat is best remembered for his work in number theory, in particular for Fermat's last theorem. This theorem states that the equation

$$x^n + y^n = z^n$$

has no nonzero integer solutions x, y and z when the integer exponent $n > 2$. Fermat wrote, in the margin of Bachet's translation of Diophantus's *Arithmetica*,

> I have discovered a truly remarkable proof which this margin is too small to contain.

These marginal notes became known only after Fermat's death, when his son Samuel published an edition of Bachet's translation of Diophantus's *Arithmetica* with his father's notes in 1670.

It is now believed that Fermat's "proof" was wrong, although it is impossible to be completely certain. The truth of Fermat's assertion was proved in June 1993, by the British mathematician Andrew Wiles at Princeton University, but Wiles withdrew the claim when problems emerged later in 1993. In November, 1994 Wiles again claimed to have a correct proof (joint with his student Richard Taylor) which has now been accepted. Unsuccessful attempts to prove the theorem over a 300-year period led to the discovery of commutative ring theory and a wealth of other mathematical developments.

Fermat's correspondence with the Paris mathematicians restarted in 1654 when Blaise Pascal wrote to him to ask for confirmation about his ideas on probability. Blaise Pascal knew of Fermat through his father, who had died 3 years earlier, and was well aware of Fermat's outstanding mathematical abilities. Their short correspondence set up the theory of probability, and from this they are now regarded as joint founders of the subject.

It was Fermat's habit to solve problems and then pose them (without proof or solution) to the community of mathematicians. Some of these were quite deep and difficult, and people found them aggravating. One problem that he posed was that the sum of two cubes cannot be a cube (a special case of Fermat's last theorem which may indicate that by this time Fermat had realized that his proof of the general result was incorrect), that there are exactly two integer solutions of $x^2 + 4 = y^3$, and that the equation $x^2 + 2 = y^3$ has only one integer solution. He posed certain problems directly to the British—just as a cross-cultural challenge. The mathematical community failed to see that Fermat had been hoping that his specific problems would lead them to discover, as he had done, deeper theoretical results.

Fermat has been described by some historical scholars as

> Secretive and taciturn, he did not like to talk about himself and was loathe to reveal too much about his thinking. . . . His thought, however original or novel, operated within a range of possibilities limited by that time [1600–1650] and that place [France].

Fermat's last theorem (or last problem) is one of the grand oldies of mathematics. It has been considered a problem of central and lasting importance ever since Fermat wrote his famous marginal comment. Many an excellent mathematician—from Sophie Germain (1776–1831) to Leonhard Euler (1707–1783) to Emil Artin (1898–1962)—worked on the problem and obtained important partial results. It was a tremendous event when Andrew Wiles announced that he had a proof of Fermat's last theorem. In fact he did so in a way that was uncharacteristically melodramatic for a mathematician.

The Isaac Newton Institute for Mathematics was constructed at Cambridge University in 1992. Isaac Newton had been a professor at Cambridge for most of his professional scholarly life, and held the Lucasian Chair of Mathematics. In fact, the oft-told Cinderella story (which is verifiably true) is that Newton's teacher, Isaac Barrow, stepped down from that position so that Newton could occupy it.[116] An extraordinary act of scholarly charity. Newton was the greatest scientist, and one of the three greatest mathematicians, who ever lived. He is remembered with reverence by the British, especially by British intellectuals. So it is natural that they would build a mathematics institute in his honor.

Among the activities at the Newton Institute are a variety of mathematics conferences. In 1993 there was to be a conference on algebraic number theory. Andrew Wiles was one of the most prominent and accomplished number theorists in the world. He had been extraordinarily quiet for several years, neither producing any papers nor giving many talks. But his reputation was still strong, and he was invited to give one of the plenary talks at this conference.

The first unusual thing that Wiles did was write to the organizers to say that he had some important and substantial things to say, and he couldn't fit his message into a single lecture. He needed three lectures to exposit his ideas. Such *chutzpah* is almost never seen in mathematics, and never from such a quiet and polite individual as Andrew Wiles. But the organizers were great admirers of Wiles's earlier work, and they acceded to his request.

It might be mentioned that, up to this point in time, Wiles had been quite secretive about his work on Fermat's last theorem. He felt that what he had been doing was rather daring—definitely off the beaten path—and he did not want to start a spate of gossip. He also, quite frankly, did not want other people horning in on his work. Perhaps more importantly, he just wanted to be left alone. Even after he had a manuscript written up, he would not show the work in its entirety (and its entirety was about 200 manuscript pages!) to even his most trusted colleagues and associates. He showed different parts to different people, and he asked for their criticism and feedback.

When the appointed time came, and Wiles delivered his thoughts to a packed audience at the Newton Institute, there was a definite sense that something exciting was happening. Of course Wiles had not told anyone the details of what he was up to, or what his main message was going to be. But, as he proceeded from the first lecture, to the second lecture, to the third lecture (on successive days), the tension and excitement were definitely building. By the third lecture it was clear that something new had been added, for there were a number of members of the press in the room. Clearly Andrew Wiles was going to do something seminal, perhaps of historical significance.

And indeed, as he wound down to the end of his third lecture, Andrew Wiles wrote on the board that a certain elliptic curve was modular. Then he drew a double arrow and wrote

---

[116]But there were other factors at play in Barrow's life. He had been offered a position at court, and he needed to resign his professorship in order to accept it.

It may be noted that Stephen Hawking occupies the Lucasian Chair today. Rare book collector John Fry owns a first edition of Newton's *Principia* that has been autographed by Hawking.

FLT. This last is number theorists' slang for "Fermat's last theorem." Andrew Wiles had just announced to the world that Fermat's last theorem was proved.

This was a huge event. The news instantly went out around the world by email. Soon everyone knew that Andrew Wiles had proved Fermat's last theorem. We have already mentioned that Wiles had written almost no papers and given almost no talks for the preceding seven years. He also had cut off communication with most of his professional friends and associates. The reason was now clear: He had been up in the attic of his house, sitting at an old desk and working on Fermat's last theorem.

There was a huge champagne celebration at the Newton Institute, and everyone hastened to tell Wiles that he was a hail fellow well met. It should be borne in mind that Wiles was at this time a number theorist of the highest reputation. If he announced something to be true, then it was felt that it must be true. Mathematicians of this caliber do not make mistakes.

At this point perhaps a little further background should be provided. Wiles had, rather cautiously, been sharing some of his ideas with his colleagues (and former distinguished colleagues) at Princeton—notably Nick Katz, Peter Sarnak, and Gerd Faltings. In fact, he had given each of these mathematicians *portions* of his manuscript to check. Nobody got to see the whole thing! And they dutifully slogged their way through the dense mathematics—much of it very original and unfamiliar—in order to aid Wiles with his work. Thus, when Wiles made his announcement at the Newton Institute, he could be fairly confident that what he was saying was correct and verifiable.

The conference in Cambridge lasted about one week, and then Wiles returned to Princeton. But now the cat was out of the bag, and pandemonium reigned in Princeton University's math department. It is safe to say that no event of this magnitude had ever taken place in Fine Hall. Breakthroughs in pure mathematics tend to be of little interest to nonspecialists. For the most part, they are too technical and recondite for anyone but another mathematician to understand. But Fermat's last theorem was different. Firstly, *anyone* could understand the problem. Secondly, the problem was over 300 years old and Fermat had posed it in such a charming and eccentric way.

Wiles's achievement seemed, in the public mind, to represent the pinnacle of the (scholarly) single-combat warrior defeating the giant Goliath of mathematics. Fine Hall was awash with reporters and news cameras and press conferences.[117] Ordinarily taciturn and painfully shy mathematicians found themselves before the camera making demure statements about the value and beauty of Wiles's work. Of course Wiles was in seventh heaven. Fermat's last theorem had been his pet problem since childhood. By solving the problem, he had fulfilled a lifelong ambition.[118]

---

[117] Andrew Wiles was even offered the opportunity to appear in a *Gap* commercial. Barbara Walters wanted to have him on her show. When Wiles revealed that he had no idea who Walters was (nor did he much care), she chose Clint Eastwood instead.

[118] An interesting sidenote on all the publicity that accompanied Wiles's accomplishment is this. The mathematicians at Princeton are no wilting flowers. They all have a pretty good opinion of themselves. And they started to feel that if Andy Wiles deserved all this publicity then perhaps they did as well. Before long, other Princeton faculty were announcing that *they too* had solved famous open problems—including the **P/NP** problem discussed above. Unfortunately none of these claims ever panned out.

Wiles at this stage participated in the ordinary process of scholarly discourse. That is, he circulated copies of his (quite long) paper all over the world. And you can bet that people *read it*. After all, this was the event of the season. The work was extremely technical and difficult to plow through, but good number theorists could run seminars and work their way through all the delicate argument and calculations.

And then disaster struck. Professors Katz and Illusie had been checking a certain portion of Wiles's argument, and they found an error. A very serious error. Wiles had claimed (without proof!) that a certain group was finite. But in fact there was no way to confirm this assertion.[119] Thus Wiles's celebrated proof of *FLT* had developed a leak. Now he was a lone hero of a new stripe. For nobody was going to step up to the plate and fix Wiles's theorem *for* him. *He* had to do it.

For a while the fact of the gap in Wiles's proof was kept "in house." Nobody wanted to embarrass Wiles. And there was definitely hope that he would be able to fix it. Now it should be noted that Wiles's teacher in England had been John Coates. Over time, a certain amount of polite professional rivalry had developed between the two. When Coates got wind of the problems with Wiles's proof of Fermat's last theorem, he started to circulate the news. Thus the cat was finally let out of the bag.

Once again, the story of Andrew Wiles emphasizes the traditional "single-combat warrior" view of mathematics that has been prevalent for over 2000 years: A single individual comes up with the ideas, a single individual writes the paper, a single individual defends it. The mathematical community at large studies the work and decides whether it is worthwhile and correct. If the work passes muster, then that single individual reaps the benefits and rewards. If the work is found to have an error, then it is the responsibility of that single point person to fix it.

Princeton University has extraordinary resources, and an extraordinary dedication to its faculty and their work. The institution gave Wiles a year off, with no duties, so that he could endeavor to fix his proof. Poor Wiles. He had made such an extravagant and bold claim, and received *such* notoriety. His picture had appeared on the front page of every major newspaper in the world. And now his program was at serious risk. Unless he could fix it.

Andrew Wiles is a devoted scholar. He believes in his mathematics, and he believes in himself, and he believed in his proof of Fermat's last theorem. He set himself to work on patching up his now-famous-but-incorrect proof. Since he did not know how to show that that particular group was finite, he developed other approaches to the problem. Time after

---

[119]There is an interesting epistemological point here that is well worth pondering. Most of the interesting errors in the mathematical literature—and this includes the present author's own work—occur when the writer says "Obviously ... " or "Clearly ... " There is no subterfuge here. The mathematician in question has been completely immersed in, and thoroughly absorbed by, this subject area for several years. He is as comfortable with the ideas as he is with his tongue in his mouth (to paraphrase a favorite statement of John E. Littlewood). So lots of things are perfectly clear to him. When the mathematician is writing up a very long, tedious, recondite proof, fatigue sets in. There will be junctures where it just seems too much to have to write out the details of a particular assertion. So comes indulgence in the conceit of saying, "Obviously ... " The mathematician is in effect saying, "If you have read this far in my paper, then you understand pretty well what is going on here. So you should be able to figure this out for yourself. Have at it." And that is how trouble comes about.

time, they all failed. As the year drew to a close, Wiles was on the brink of despair. But then he had a stroke of luck.

Wiles had been sharing his thoughts with his graduate student Richard Taylor. Taylor was extremely gifted (he is now a professor at Harvard), and could understand all the intricacies of Wiles's approach to the *FLT*. He recommended that Wiles return to his original approach to the problem, and he contributed some ideas of his own on how to deal with the difficulty of that group being finite. Thus, together Wiles and Taylor retackled the original program that Wiles had set for proving Fermat's last theorem. And finally they succeeded. The benchmark was that they gave Fields Medalist Gerd Faltings the *entire new manuscript*, and he read it straight through. At the end Faltings declared, "Wiles and Taylor have done it. *FLT* is proved." And that was it. This is how the mathematical method operates in our world. One good mathematician asserts that he can prove a new theorem. He writes down a proof in the accepted argot and shows this recorded proof to one or more expert colleagues. Then they assess the work and pass judgment. There is now an entire issue of the *Annals of Mathematics*—in volume 141, 1995—containing the proof of *FLT*. This consists of a long paper by Wiles alone and a somewhat shorter paper by Taylor and Wiles. Although Wiles generally receives the lion's share of the credit for this great accomplishment, Taylor's contribution must be recorded as being key to the solution.

One of the reasons that *mathematicians* were so excited by Wiles's work on Fermat's last theorem is that it built on so many of the important ideas of number theory that had been developed in the preceding fifty years. One of the keys to what Wiles did was Ken Ribet's (Professor of Mathematics at U.C. Berkeley) result that the famous Taniyama–Shimura–Weil Conjecture implies *FLT*. For many years Wiles had kept his lifelong interest in Fermat's last theorem something of a secret. It was considered to be one of those impossibly difficult problems—300 years old and something that was perhaps best worked around—and he would have been embarrassed to let people know that he was spending time on it. But Ribet's result put *FLT* back in the mainstream of number theory. It made *FLT* a matter of current interest. It hooked up the great unsolved problem with key ideas in the main flow of modern number-theoretic research.

The key upshot of the trail of thought in the last paragraph is that if Wiles could prove the Taniyama–Shimura–Weil Conjecture, then Fermat's last theorem would follow. In the end, Wiles did not succeed in doing that. Instead he ended up proving a *modified* or limited version of Taniyama–Shimura–Weil, and that was sufficient to prove *FLT*.

By today a great many people have checked Andrew Wiles's proof. There have been simplifications to some of the steps, and the problem has also been generalized (the famous "*ABC* conjecture" is one such generalization that has become quite famous). Fermat's last theorem is now considered to be a genuine theorem, and Wiles's ideas have become an important part of the fabric of number theory.

Wiles's work has placed the spotlight on all the ideas that contributed to his victory over Fermat's last theorem. In particular, the Taniyama–Shimura–Weil conjecture shared some of the limelight. As a result, that conjecture has now been fully proved as well. This is an elegant example of how mathematics works in practice.

11.9 The Wily Infinitesimal

When calculus was first invented by Newton and Leibniz, people were not quick to embrace these new ideas. The techniques of calculus were evidently powerful, and could be used to solve a host of problems that were long considered to have been intractable. To be sure, Newton successfully studied and analyzed

- motion
- gravity
- refraction of light
- the motions of the planets
- mechanics

But there were a number of theoretical underpinnings of calculus that were not well understood—even by the subject's inventors. Among them was the concept of limit—which actually would not be properly understood until Augustin-Louis Cauchy gave a rigorous definition in 1821. But the real sticking point in the subject was infinitesimals. An infinitesimal to Isaac Newton was a number that was positive, but smaller than any ordinary positive real number. Thus an infinitesimal is a *positive* number that is smaller than $1/10$, smaller than $1/1000$, smaller than 10^{-10}, smaller than the radius of a proton. How could such a quantity exist?

No less an eminence than Bishop Berkeley (1685–1753)—who was himself a considerable scholar—weighed in with a skeptical view of the calculus. One of Berkeley's broadsides reads:

> All these points, I say, are supposed and believed by certain rigorous exactors of evidence in religion, men who pretend to believe no further than they can see. That men who have been conversant only about clear points should with difficulty admit obscure ones might not seem altogether unaccountable. But he who can digest [infinitesimals] need not, methinks, be squeamish about any point of divinity.

The Bishop succeeded in casting a pall over calculus that was not to be lifted for nearly two hundred years.

The truth is that nobody—not even Newton himself—really knew what an infinitesimal was. For Newton infinitesimals were a terrific convenience—they enabled him to perform incisive physical and mathematical deductions that led to daring and seemingly correct conclusions. Yet the foundations for what he was doing seemed to be suspect.

Cauchy's theory of limits, which came much later, showed us how to "work around" infinitesimals. But the infinitesimals were still there, and physicists and others loved to reason using infinitesimals—they were compelling and heuristically appealing, and led to useful and powerful conclusions. But nobody could say what they were. Then Abraham Robinson came to the rescue.

Robinson was a mathematician remarkable for his breadth and depth. He studied both *very* applied problems and *very* abstract and pure problems. He was universally admired and respected for his many and varied contributions to mathematics. In 1963 Robinson created a new subject area called "nonstandard analysis." The most important feature of this new

the field.[121] For the time being, the subject of this big discovery will be the most important thing in sight. And, as long as significant new ideas are forthcoming, the area will be hot. But after a while it will fade from view. It can be revived if someone comes along with a new idea, and this often happens. But just as frequently it does not.

So if your new paper, the product of your intense labor over a protracted period of time, is in an area that is of broad interest, then you will find that you will earn much appreciative commentary, you will be invited to speak at conferences, and you will enjoy some of the warmest encomia of the profession. In the other case your paper will be published, put on the shelf, and more or less forgotten.

What we have just described is the ordinary course of life, not just for the mathematician but for academics in general and for people who are engaged in any line of creative work—whether it be music or painting or dance or cooking. A mature mathematician realizes that working in mathematics satisfies a personal intellectual curiosity and desire to work on one's own ideas. For most of us that is more than enough reason to continue to pursue the holy grail of the next theorem.

Of course, as we have indicated elsewhere, it also can happen that a new paper has a mistake in it. If it is a small mistake, then it is likely you can fix it and press on. If it is a big mistake, then you may have to shelve the paper and find another pursuit to justify your existence. It is occasionally possible to rectify even this dire situation. Andrew Wiles made a big mistake in his first proof of Fermat's last theorem (see Section 11.8). It took him a full year to fix it, but he finally did. And then he reaped the rewards.

11.10.1 Frustration and Misunderstanding

Probably the most frustrating, and often unfixable, situation is that in which you are quite sure that your new paper is correct but the world does not accept it. That is, you believe that you have given a bona fide proof of a significant new result, but the community of mathematicians does not believe your proof. If you are in such a situation, and you are lucky, someone will sit down with you and show you where your error is. But many times the world just ignores a paper that it does not believe, and simply forges ahead.

And where does that leave you? You can go on the lecture circuit and try to convince people that you really have the goods. But the fact is that mathematics is vetted and verified in a very special and precise way: individual mathematicians sit down and read your work. They either believe it or they do not. And in the latter case you are left in an extremely awkward situation.

We have provided this extensive introduction to set the stage for now listing a few famous instances of people who have written significant papers or books purporting to solve important problems, but who have found that the mathematics profession has not been willing to validate and accept the work. In most of these cases the progenitors have experienced bitterness, unhappiness, and frustration. In some instances the experience changed the

---

[121] As an instance, in 2006 Terence Tao of UCLA won the Fields Medal. He has had a great many important new ideas. Now UCLA is a good school, but it only ranks #26 in the *U.S. News and World Report* college ranking (#12 in the math graduate program ranking). Great ideas and brilliant minds can reside anywhere.

direction of a person's career. In other instances, the person quit doing mathematics altogether.

- In 1969 Alfred Adler of the State University of New York at Stony Brook published a paper in the *American Journal of Mathematics* [ADL1] purporting to show that there is no complex manifold structure on the six-dimensional sphere. Complex manifolds are surfaces that have certain important algebraic and analytic properties. These play a significant role in differential geometry and mathematical physics. But it is difficult to come up with concrete examples of complex manifolds. The six-dimensional sphere would be an interesting and significant example if indeed it had a complex structure. Adler asserted that it did not. And he got his paper published in a top-ranked journal, so one may infer that a solid referee with strict standards agreed.

  But people have not accepted Adler's proof. For 39 years Adler has maintained that his proof is correct. He is now retired, and no longer participating in the discourse. But he produced no further published mathematical work after his 1969 paper.[122] And, for a good many of those 39 years, nobody was able to put their finger on where the mistake was. Finally, about 5 years ago, Yum-Tong Siu of Harvard wrote a paper in which he explained precisely where the error was. But Adler's paper has never been repaired, and nobody knows to this day how to prove the assertion either true or false.

- In 1993 Wu-Yi Hsiang published his solution of the Kepler sphere-packing problem (see the detailed discussion in Section 10.3). He published the paper in a journal of which he was an editor, and perhaps he received some friendly treatment from the journal. That is, the paper may not have been subjected to the most rigorous refereeing. Even before the publication of the paper, the ranking experts had mounted some objections to Hsiang's arguments. He had claimed in the paper that he had identified the "worst case scenario" for sphere packing, and he then proceeded to analyze that particular case. The experts did not find this argument to be either complete or compelling. But Hsiang believed in his methodology and determined to publish the work. And there it stands. Hsiang's methodology was traditional mathematics—he used methods of spherical trigonometry. Anyone who chooses to do so can read the paper and check the work. The experts did so, and in the end did not accept it as a proof. Hsiang still stands by his ideas; in 2001 he published a book [HSI3] explaining his approach to the problem.

  Meanwhile, Thomas Hales (again see Section 10.3) has produced a new proof of the Kepler sphere-packing conjecture. It has been published in the prestigious *Annals of Mathematics* [HAL2], and was subjected to rigorous refereeing over a long period of time. But Hales's arguments rely heavily on intense and protracted computer calculations which cannot possibly be checked by a human being. There are now techniques for having a second computer check the work of the first computer, and Hales has organized a project to carry out such a verification. He anticipates that it be finished and ready for publication in late 2011.

---

[122] Alfred Adler did, in 1972, publish a remarkable article [ADL2] in the *New Yorker* describing what it is like to be a mathematician. It is rather a bittersweet article, exhibiting considerable frustration with the profession. One can only surmise that the provenance of this article is Adler's negative experience with that 1969 paper.

- In 1978 Wilhelm Klingenberg published a paper, followed by two books (see [KLI1]–[KLI3]), in which he asserted that any closed, bounded surface in space has infinitely many distinct closed geodesics. Here a *geodesic* is a curve of least length. For example, in the plane, the curve of least length connecting two points is a straight line. On a sphere (like the earth), the curve of least length connecting two points is an arc of a great circle (here a *great circle* is the circular arc that obtains when a plane through the center of the sphere intersects the sphere). Airlines follow great circle paths when flying from Los Angeles to Paris, for example. Klingenberg is a distinguished mathematician who had spent his entire career studying geodesics. He had made many important contributions, and was the leader in the field. But people did not accept his proof.

  The problem had a long history. The distinguished G.D. Birkhoff, around 1914, proved the existence of at least one geodesic for any metric on a closed sphere. Lyusternik and Schnirelmann proved twenty years later that the sphere in 3-space, equipped with any Riemannian metric, has at least three geodesics. It was also proved around that time that any closed, bounded surface (equipped with any metric) has at least one closed geodesic. Later, in the 1960s and 1970s, Gromoll and Meyer and Sullivan showed that a surface with sufficiently complicated topology would have infinitely many distinct geodesics. Thus they left open the problem for surfaces with simple geometry/topology. Then came Klingenberg. He claimed, using profound ideas from Morse theory, that he could solve the whole problem. But he had trouble convincing the world.

  Mathematician Friedrich Hirzebruch used to hold yearly conferences in Bonn that were dubbed the *Arbeitstagungen*. These were quite unusual events. The first day of the conference was spent with *all* the participants debating who should be allowed to speak and why. After Klingenberg announced his result, he was asked to speak at the Arbeitstagung. But the audience really gave him a hard time, peppered him with questions, and cast doubt upon various parts of his proof. To this day there is no consensus as to whether Klingenberg's proof of this important result is correct. And there has been no meaningful progress in fixing the matter, although Bangert and Franks have come up with a new approach to the question. They have a result on the 2-sphere in 3-space.

- We have already told the story, in Section 10.4, of William P. Thurston's geometrization program. This is surely one of the dazzling new ideas in geometric topology, and it would imply the celebrated Poincaré conjecture. Thurston was, in the end, able to produce a proof of the program for Haken manifolds and for some other special cases. A complete proof of the geometrization program, from its progenitor's point of view, has never appeared. Although there are many who believe, both on a priori grounds and because they are sold on Thurston's outline proof, that Thurston has a bona fide theorem, the world at large is still waiting for the details. Grisha Perelman's new arguments (see Section 10.5) may settle the matter once and for all. But it will be some time before we know this for certain.

- When Frederick Almgren died in 1997, he left behind a 1728-page manuscript purporting to prove an important regularity theorem for minimal surfaces. These are surfaces that model soap films, polymers, and other important artifacts of nature. Now it should be stressed that Almgren was a very distinguished mathematician, a pioneer in his field, and a professor at Princeton University. On a priori grounds, people are inclined to believe Almgren's theorem. But it is essentially impossible for any human being to digest the 1728 pages of dense and technical mathematics that Almgren produced. Two of his students, Vladimir Scheffer and Jean Taylor (also Almgren's wife) published Almgren's manuscript posthumously as a book [ALM]. But it may be a long time before this work has gone through the usual vetting and has been given the official imprimatur.

• We conclude this section by relating the story of a proof that caused a huge fight to occur. The incident is more than fifty years old, yet there remain bitter feelings. The two chief participants are very distinguished mathematicians. But everyone took a side in this matter, and the raw feelings still fester. The issue here is *not* whether a given proof is correct. Rather, the question is whose proof it is.

In 1948 Paul Erdős was in residence at the Institute for Advanced Study and Atle Selberg (1917–2007) was a permanent member. Selberg got a promising idea for obtaining an elementary proof of the prime number theorem (that is, a proof that does not use complex analysis). Selberg only lacked a particular lemma to complete his argument. See Section 11.1 for background.

The next day, Paul Erdős was able to supply the lemma that Selberg needed. Later, Selberg was able to modify his proof so that it did not require Erdős's idea. Unfortunately, a terrible priority dispute erupted between Erdős and Selberg. Since he was able to contribute an important step to the argument, Erdős just assumed that he and Selberg would write a joint paper on the result. Selberg, who had never written a joint paper with anyone, had other plans.

One version of the tale is that Selberg was visiting at another university, sitting in somebody's office and having a chat. Another mathematician walked in and said, "I just got a postcard saying that Erdős and some Norwegian guy that I never heard of have found an elementary proof of the prime number theorem." This really set off Selberg, and he was then determined to write up the result all by himself.

Dorian Goldfeld (1947–) has taken pains to interview all survivors who participated in or witnessed the feud between Erdős and Selberg. It is clear that nobody was wrong and nobody was right. Both Erdős and Selberg contributed to this important discovery, but they had a significant clash of egos and of styles. Irving Kaplansky (1917–2006) was in residence at the Institute for Advanced Study in those days and witnessed the feud in some detail. He tells me that at one point he went to Erdős and said, "Paul, you always say that mathematics is part of the public trust. Nobody owns the theorems. They are out there for all to learn and to develop. So why do you continue this feud with Selberg? Why don't you just let it go?" Erdős's reply was, "Ah, but this is the prime number theorem."

Today, when the elementary proof of the prime number theorem is mentioned (either in books or in articles or in conversation), both Erdős and Selberg are named. People try to avoid pointing the finger or taking sides. But some dreadful misinformation sometimes slips out. In the biography [HOF] of Erdős, the author claims that Selberg "stole" the Field Medal from Erdős. This is preposterous. Selberg deserved the Fields Medal for lots of reasons, and would have won it whether Erdős were ever in the picture or not. [Erdős was a great mathematician, but never a likely Fields Medalist.]

12

John Horgan and "The Death of Proof?"

If, as they say, some dust thrown in my eyes
Will keep my talk from getting overwise,
I'm not the one for putting off the proof.
Let it be overwhelming.

—Robert Frost

We are talking here about theoretical physics, and therefore of course mathematical rigor is irrelevant and impossible.

—Edmund Landau

The folly of mistaking a paradox for a discovery, a metaphor for a proof, a torrent of verbiage for a spring of capital truths, and oneself for an oracle, is inborn in us.

—Paul Valéry

Ah, Why, ye Gods, should two and two make four?

—Alexander Pope

To foresee the future of mathematics, the true method is to study its history and its present state.

—Henri Poincaré

12.1   Horgan's Thesis

In 1993 John Horgan, a staff writer for *Scientific American*, published an article called *The Death of Proof?* [HOR1]. In this piece the author claimed that mathematical proof no longer had a valid role in modern thinking. There were several components of his argument, and they are well worth considering here.

First, Horgan was a student of literary critic Harold Bloom at Yale University. One of Bloom's principal theses—in the context of *literature*—is that there is nothing new under the sun. Any important and much-praised piece of modern literature is derivative from one of the classics by Shakespeare or Chaucer or Spencer or Dryden or some other great figure from 400 years ago. Horgan endeavored to shoehorn modern scientific work into a similar mold. He claimed that Isaac Newton and the other great minds of 300 years ago had all the great ideas, and all we are doing now is turning the crank and producing derivative thoughts.

A second vector in Horgan's argument is that computers can do much more effectively what human beings have done traditionally—which is to *think*. Put in slightly different words, a mathematical dinosaur might claim that the deep insights of mathematics can be discovered and produced and verified (i.e., proved) only by a human being with traditional training in mathematics. John Horgan would claim that a computer can do it better and more effectively and much faster.

The third component of John Horgan's thought was that mathematical proofs have become so complicated (witness Wiles's proof of Fermat's last theorem, which takes up an entire issue of *Annals of Mathematics*) that nobody can understand them anyway. So how could they possibly be playing any significant role in the development of modern mathematical science?

The trouble with Horgan's arguments is that he is thinking like a literary critic, and reasoning by analogy. One may agree or disagree with Harold Bloom's thesis (for instance, from what classical work is Alan Ginsburg's *Howl* derivative? Or James Joyce's *Finnegans Wake*?). But there is no evidence that it applies to modern scientific progress. Isaac Newton, to take just one example, was a great thinker. He created the modern scientific method, he created mathematical physics, he (along with Leibniz) invented calculus, and created the most powerful body of scientific/analytic tools ever devised. There may never be another scholar to equal Newton. But each generation of scientific work builds atop earlier work. It would be quite difficult to argue that relativity theory is derivative from the work of 300 years ago. Or that quantum theory is derivative from ideas of Maxwell. Or that string theory is derivative from ideas of Fermat.

Moreover, while computers are terrific at manipulating data accurately and rapidly, computers cannot think—at least not in the way that humans think. Artificial intelligence software is an effort to make the computer perform tasks that the human brain can perform. Some notable successes have been achieved, but they are rather elementary. They don't begin to approach or emulate the power and depth of something like Wiles's proof of Fermat's last theorem.

We cannot help but agree with Horgan that mathematical proofs have become rather complicated. Two hundred years ago, published mathematics papers tended to be fairly short, and arguments were rarely more than a few pages. Today we are considerably more sophisticated, mathematics is a much larger and more complex and recondite enterprise, and yes, indeed, many mathematics papers run to fifty or more pages and have enormously complex proofs. But it is incorrect to assert that other mathematicians are helpless to check them. In fact they do. It is an enormous amount of work to do so, but if a result is important, and if the proof introduces significant new techniques, then people will put in the time to study the work and validate it. Many people have studied Wiles's proof, and it is quite certain that it is correct. Moreover, people have simplified several parts of the proof. And there are important generalizations of Fermat's last theorem, such as the ABC conjecture, which put the entire problem into a new perspective and will no doubt lead to further simplifications.

Mathematics is a process, marked by milestones such as solutions of the great problems and discoveries of marvelous new theorems. But over time, the big, complicated ideas get digested and worked over and built into the infra-structure of the subject. In the end they are simplified and rendered natural and understandable. Gauss actually anticipated by many years the discovery (by Bolyai and Lobachevsky) of non-Euclidean geometry. He did not share or publish his ideas because he felt that they were too subtle; nobody would understand them. Today we teach non-Euclidean geometry to high school students. The Cauchy integral, Galois theory, the Lebesgue integral, cohomology theory, and many other key ideas of modern mathematics were considered to be impossibly complex and erudite when they were first discovered. Today we teach them to undergraduates.

John Horgan's article created quite a stir. In the mathematics community, this author became the point man for responding to the piece; the rebuttal appears in [KRA1]. In my article, I make the points just enunciated, but in considerably more detail. It is safe to say that the mathematics community has evaluated Horgan's arguments and rejected them.

But Horgan in fact leveraged his thesis into something more substantial. He subsequently published a book called *The End of Science* [HOR2]. This work carried on many of the themes introduced in the *Scientific American* article. In his book the author claims that scientific research has come to an end. All the great ideas were discovered prior to 1900 and what we are doing now is a sham—just a way of scamming National Science Foundation grant money from the government.

This new claim by Horgan is reminiscent of repeated efforts in the nineteenth and early twentieth centuries to close the U. S. Patent Office—with the claim that everything had already been invented and there was nothing more to do. Even in the past 35–40 years the number of new inventions has been astonishing—ranging from the personal computer to the cellular telephone to the Segway transportation device. Again, Horgan is guilty in this book of arguing by analogy. He advocates, for instance, thinking about how it looks when you are driving a car into a brick wall. In the last few moments, it appears as though the wall is approaching ever faster. Just so, Horgan reasons, it looks as though science is making progress at an exponentially increasing rate because in fact it is running into a brick wall (of no progress).

It is difficult to take such arguments seriously. Reasoning like this, which may be most apposite in the context of literary theory, make little sense in the milieu of science. It seems irrefutable that science is in a golden age. The genome project has opened up many new doors. String theory is really changing the face of physics. There are so many new directions in engineering—including biomedical engineering—that it would be impossible to describe them all. Chemistry changes our lives every day with new polymers and new synthetic products.

12.2 Will "Proof" Remain the Benchmark for Mathematical Progress?

The great advantage of a traditional mathematical proof is that it is a bulletproof means of verifying and confirming an assertion. A statement that is proved today by the methods introduced by Euclid in his *Elements* will still be valid 1000 years from now. Changing fashions, changing values, and changing goals do not affect the validity of a mathematical proof. New discoveries do not invalidate old ones when the old ones were established by proof.

For this reason that mathematical proof will continue to be a standard to which we can all aspire. But we can at the same time tolerate and appreciate other approaches to the matter. Mathematical proofs do not traditionally carry a great deal of weight in a physics or engineering department. Heuristic reasoning traditionally does not hold much water in a mathematics department. As indicated in the beginning of this book, a proof is a psychological device for convincing somebody that something is true. If that "somebody" is a classically trained mathematician, then most likely the proof that is desirable is a traditional one a long, precise computer calculation may be more convincing to such a person. Euclid-style, logical argument. If instead that "somebody" is a modern numerical analyst, then he/she may be more convinced by a long and very precise computer calculation.

Note that neither of the two scenarios painted at the end of the last paragraph supplants the other. They serve two different purposes; their intent is to bring two different types of people on board to a certain program. In fact, properly viewed, they complement each other.

One of the main points of the present book is that "proof"—the traditional, Euclid-inspired, razor-sharp arrow of argumentation leading inexorably to a precise conclusion—is immortal. It was invented to serve a very specific purpose, and it does so admirably. But this mode of thought, and of verifying assertions, is now put into a new context. It is a very robust, and supportive, and stimulating context that only *adds* to what we can do. It augments our understanding. It increases our arsenal of weapons. It makes us all stronger.

13

Closing Thoughts

Physics has provided mathematics with many fine suggestions and new initiatives, but mathematics does not need to copy the style of experimental physics. Mathematics rests on proof—and proof is eternal.

—Saunders Mac Lane

Mathematics seeks to reduce complexity to a manageable level and also to impose structure where no structure is apparent.

—Michael Aschbacher

All that is missing is a proof.

—John Milnor

Never try to teach a pig to sing. It frustrates you and irritates the pig.

—Anonymous

Life is good for only two things, discovering mathematics and teaching mathematics.

—Siméon Poisson

. . . it is impossible to write out a very long and complicated argument without error, so is such a "proof" really a proof?

—Michael Aschbacher

13.1   Why Proofs Are Important

Before proofs, about 2600 years ago, mathematics was a heuristic and phenomenological subject. Spurred largely (though not entirely) by practical considerations of land surveying, commerce, and counting, there seemed to be no real need for any kind of theory or rigor. It was only with the advent of abstract mathematics—or mathematics for its own sake—that it began to become clear why proofs are important. Indeed, proofs are central to the way we now view our discipline.

There are tens of thousands of mathematicians all over the world today. Just as an instance, the *Notices of the American Mathematical Society* has a circulation of about 30,000.

(This is the news organ, and the journal of record, for the American Mathematical Society.) And the actual readership is larger still. And abstract mathematics is a well-established discipline. There are few with any advanced knowledge of mathematics who would argue that proof no longer has a place in this subject. Proof is at the heart of the discipline; it is what makes mathematics tick. Just as hand–eye coordination is at the heart of hitting a baseball, and practical technical insight is at the heart of being an engineer, and a sense of color and aesthetics is at the heart of being a painter, so an ability to appreciate and to create proofs is at the heart of being a mathematician.

If one were to remove "proof" from mathematics, then all that would remain would be a descriptive language. We could examine right triangles, and congruences, parallel lines, and attempt to learn something. We could look at pictures of fractals and make descriptive remarks. We could generate computer printouts and offer witty observations. We could let the computer crank out reams of numerical data and attempt to evaluate those data. We could post beautiful computer graphics and endeavor to assess them. *But we would not be doing mathematics.* Mathematics is (i) coming up with new ideas and (ii) validating those ideas by way of proof.[123] The timelessness and intrinsic value of the subject come from the methodology, and that methodology is proof.

Again we must emphasize that what sets "proof" apart from the methodologies of other disciplines is its timelessness. A great idea in computer science could easily be rendered "old school" in a couple of years. The computer languages SNOBOL and COBOL were hot in the 1960s, but hardly anyone uses them anymore. In the 1970s, Fortran was the definitive computer language for scientific computation. Today it has been superseded by C and C++, in the sense that the latter two languages tend to be the default. But modern versions of Fortran such as Fortran-90, Fortran-95, and Fortran-2000 are still in general use.

And so it is in medicine. You may recall that radial keratotomy was for a short time the hottest thing around for correcting vision through surgery. This lasted just a couple of years, until medical scientists realized that they could not accurately predict the long-term effects of the procedure. It has been replaced by laser-assisted *in situ* keratomileusis or LASIK surgery. Now there are questions about the long-term effects of LASIK, so that methodology is being reevaluated.

Artificial hearts were developed because patients were rejecting transplanted hearts. But now doctors have figured out how to get patients to accept the transplants; so artificial hearts are of less interest. It used to be that X-ray was the definitive diagnostic tool. We know that X-rays are still useful, but in many applications they are replaced by magnetic resonance imaging or one of the many other new imaging technologies that have been developed.

We have described some of these events in detail to emphasize that this sort of thing *never* happens in mathematics. To be sure, mathematics has its fashions and its prejudices. But in mathematics, once correct is always correct. For a time, everyone in college studied spherical trigonometry; it was just part of the curriculum, as calculus is today. Then it fell

---

[123]Mathematics is this and so much more. This book has endeavored to portray mathematics as a multifaceted beast. In today's world, mathematics is proofs and algorithms (both proved and heuristic), theories, methodologies, approaches, conjectures, models, and many additional facets that are developing every day.

out of fashion. However, in 1993, Wu-Yi Hsiang used spherical trigonometry to attempt a solution of the venerable Kepler sphere-packing problem. So now there is renewed interest in spherical trigonometry. Hyperbolic geometry was something of a relic of classical Riemannian geometry until Bill Thurston came along in the 1980s and made it part of his program to classify all 3-manifolds. Now everybody is studying hyperbolic geometry.

But it must be emphasized that spherical trigonometry and hyperbolic geometry were never declared *wrong*. There was never any danger of that happening. It's just that the world passed these subjects by for a while. Everybody knew they were there, and what they were good for. They simply did not attract much interest. There were too many new and exciting things to spend time on. But now, because of some good ideas of some excellent mathematicians, they have been brought to the fore again.

Proofs remain important in mathematics because they are our signpost for what we can believe in, and what we can depend on. They are timeless and rigid and dependable. They are what hold the subject together, and what make it one of the glories of human thought.

13.2 Why It Is Important for Our Notion of Proof to Evolve and Change

As described in this book, our concept of what proof *is* was developed and enunciated by the ancient Greeks 2500 years ago. They set for us a remarkable and profound paradigm from which we have not wavered in the intervening millennia. But ideas change and develop. Ultimately our goal is to reach the truth, to understand that truth, to verify that truth, and to disseminate and teach that truth. For a good part of our professional history mathematicians lived inside their heads. Their primary interest was in discovering new mathematics and validating it. Sharing it with others was of secondary interest. Now our view, and our value system, has changed.

Today there are more mathematicians than there were during the entire period 500 BCE to 1950 CE. A good many mathematicians work at colleges and universities. There are more than 4100 colleges and universities (2474 four-year institutions and 1666 two-year institutions) in the United States alone, and many more thousands of institutions of higher learning around the world. In America alone there are 17.5 million students in our colleges and universities. There are also many thousands of mathematicians employed in industry, in government research facilities (such as Los Alamos and Oak Ridge), and in research laboratories of all kinds. Because mathematics is so *diverse* and *diffuse*, it has become essential that there be more communication among different types of mathematicians.

As we have described in this book, the notion of what it means to be a mathematician has grown and developed over time. Not long ago, a mathematician was an expert in Euclid's geometry, Newton's calculus, Gibbs's vector analysis, and a few other well-worn and crusty subject areas. There are many different types of mathematicians today with many different kinds of backgrounds. Some mathematicians work on the genome project. Some mathematicians work for NASA. Some work for Aerospace Corporation. Some work for the National Security Agency. Some work for financial firms on Wall Street. They often speak different languages and have different value systems. In order for mathematicians with different pedigrees to communicate effectively, we must be consciously aware of how different types of mathematical scientists approach their work. What kinds of problems do they study? What types of answers do they seek? How do they validate their work? What tools do they use?

For these reasons it is essential that the mathematical community have a formal recognition of the changing and developing nature of mathematical proof. The classical notion of proof, taught by Euclid and Pythagoras, is the bedrock of our analytical thinking procedures. Most mathematical scientists are not advocating that we abandon or repudiate the conceptual basis for our subject. What *is* true is that many different points of view, many different processes, many different types of calculation, many different sorts of evidence may contribute to the development of our thoughts. And we should be welcoming to them all. One never knows where the next idea will come from, or how it may come to fruition. Since good ideas are so precious and so hard to come by, we should not close any doors or turn away any opportunities.

Thus our notion of "proof" will develop and change. We may learn a lot from this evolution of mathematical thought, and we should. The advent of high-speed digital computers has allowed us to see things that we could never have seen before (using computer graphics and computer imaging) and to do "what if" calculations that were never before feasible. The development and proliferation of mathematical collaboration—both within and without the profession—has created new opportunities and taught us new ways to communicate. And as part of the process, we have learned to speak new languages. Learning to talk to engineers is a struggle, but one side benefit of the process is that we gain the opportunity of learning many new problems. Likewise for physics and theoretical computer science, biology, and medicine.

The great thing about going into mathematics in the twenty-first century is that it opens many doors and closes few of them. The world has become mathematized, and everyone is now conscious of this fact. People also appreciate that mathematicians have critical thinking skills, and are real problem-solvers. Law schools, medical schools, and many other postgraduate programs favor undergraduate math majors because they know that these are people who are trained to think. The ability to analyze mathematical arguments (i.e., *proofs*) and to solve mathematical problems is a talent that works in many different environments and finds applications in a number of different contexts.

13.3   What Will Be Considered a Proof in 100 Years?

It is becoming increasingly evident that the delineations among "engineer" and "mathematician" and "physicist" are becoming ever more vague. The widely proliferated collaboration among these different groups is helping to erase barriers and to open up lines of communication. Although "mathematician" has historically been a much-honored and respected profession, one that represents the pinnacle of human thought, we may now fit that model into a broader context.

It seems plausible that in 100 years we will no longer speak of mathematicians as such but rather of *mathematical scientists*. This will include traditional, pure mathematicians to be sure, but also a host of others who use mathematics for analytical purposes. It would not be at all surprising if the notion of "Department of Mathematics" at the college and university level gives way to "Division of Mathematical Sciences."

In fact, we already have a role model for this type of thinking at the California Institute of Technology (Caltech). For Caltech does not have departments at all. Instead it has divisions. There is a Division of Physical Sciences, which includes physics, mathematics, and astronomy. There is a Division of Life Sciences that includes Biology, Botany, and several other fields. The philosophy at Caltech is that departmental divisions tend to be rather artificial, and tend to cause isolation and lack of communication among people who would benefit distinctly from cross-pollination. This is just the type of symbiosis that we have been describing for mathematics in the preceding paragraphs.

So what will be considered a "proof" in the next century? There is every reason to believe that the traditional concept of pure mathematical proof will live on, and will be designated as such. But there will also be computer proofs, proofs by way of physical experiment, and proofs by way of numerical calculation. This author has participated in a project—connected with NASA's space shuttle program—that involved mathematicians, engineers, and computer scientists. The contributions from the different groups—some numerical, some analytical, some graphical—reinforced each other, and the end result was a rich tapestry of scientific effort. The end product is published in [CHE1] and [CHE2]. This type of collaboration, while rather the exception today, is likely to become ever more common as the field of applied mathematics grows, and as the need for interdisciplinary investigation proliferates.

Today many mathematics departments contain experts in computer graphics, experts in engineering problems, experts in numerical analysis, and experts in partial differential equations. These are all people who thrive on interdisciplinary work. And the role model that they present will influence those around them.

The mathematics department that is open to interdisciplinary work is one that is enriched and fulfilled in a pleasing variety of ways. Colloquium talks will cover a broad panorama of modern research. Visitors will come from a variety of backgrounds, and represent many different perspectives. Mathematicians will direct Ph.D. theses for students from engineering and physics and computer science and other disciplines as well. Conversely, mathematics students will find thesis advisors in many other departments. One already sees this happening with students studying wavelets, harmonic analysis, and numerical analysis. The trend will broaden and continue.

So the answer to the question is that "proof" will live on, but it will take on new and varied meanings. The traditional idea of proof will prosper because it will interact with other types of verification and affirmation. And other disciplines, ones that do not traditionally use mathematical proof, will come to appreciate the value of this mode of intellectual discourse.[124] The end result will be a richer tapestry of mathematical science and mathematical work. We will all benefit from the development.

---

[124] We must repeat that some of these "other disciplines" are already impressively appreciative of the mathematical method. Both the *IEEE Transactions on Signal Processing* and the *IEEE Transactions on Image Processing* contain copious mathematics, and many results that are proved.

Index of Names

Abel, Niels Henrik (1802–1829)   One of the great geniuses of 19th century mathematics. He died young of poverty and ill health. Today he is remembered by the prestigious Abel Prize.

Aiken, Howard (1900–1973)   One of the pioneers of computing. Inventor of the Harvard Mark I machine.

Albert, Adrian (1905–1972)   An outstanding algebraist of the 20th century. A faculty member at the University of Chicago. Did important cryptographic work for the military during World War II.

Artin, Emil (1898–1962)   An outstanding figure in 20th century algebra. A professor at Princeton University. Had many important doctoral students. His son Michael Artin is professor of mathematics at MIT.

Atanasoff, John (1903–1995)   A pioneer in the development of computers. Helped to invent the ABC Computer.

Atiyah, Michael (1929–)   One of the great mathematicians of the 20th century. A profound geometer, and winner of the Fields Medal and the Abel Prize.

Babbage, Charles (1791–1871)   A pioneer in the theory of computing machines. Inventor of the "analytical engine." Pioneered the idea of storing computer instructions on punch cards. Lord Byron's daughter Augusta did programming work for Babbage.

Baker, H. F. (1866–1956)   An historian of mathematics, and a particular expert on J. J. Sylvester.

Banach, Stefan (1892–1945)   One of the pioneers of functional analysis in Poland in the 1920s. A scion of the Scottish Cafe, where many famous Polish mathematicians gathered regularly. Died in a German concentration camp.

Berkeley, Bishop George (1685–1753)   A great philosopher of the 18th century. An expert in metaphysics and epistemology. Noted as a critic of Newton's calculus.

de Branges, Louis (1932–) A noted American mathematician of the 20th century. Famous for having proved the Bieberbach conjecture.

Berry, Clifford (1918–1963) One of the developers of the early ABC computer.

Birkhoff, George D. (1884–1944) One of the first great American mathematicians. Noted for having tackled Poincaré's last theorem, and also for proving the general ergodic theorem.

Bishop, Errett (1928–1983) A remarkable mathematical analyst of the 20th century. Noted for being the father of constructive analysis.

Bohr, Harald (1887–1951) Noted Dutch analyst of the 20th century. Friend of G. H. Hardy. Famous for his studies of almost periodic functions.

Bolyai, János (1802–1860) Mathematician noted for having invented non-Euclidean geometry.

Bose, Amar (1929–) Professor of Electrical Engineering at MIT. Noted for being the father of Bose Acoustics.

Brahe, Tycho (1546–1601) Teacher of Johannes Kepler and noted astronomer. Gathered the data that led to Kepler's three laws.

Brouwer, Luitzen E. J. (1881–1966) Noted 20th century topologist. Proved the Brouwer fixed-point theorem, which he later repudiated in favor of intuitionism.

Byron, Augusta (1815–1852) Daughter of Lord Byron the poet. Also a programmer for Charles Babbage's analytical engine.

Byron, Lord George Gordon (1788–1824) Noted poet.

Cantor, Georg (1845–1918) The father of set theory, and known particularly for the theory of cardinal numbers. Helped us to understand the orders of infinity.

Cartan, Élie (1869–1951) A great geometer of the 20th century. Father of the theory of differential forms. Also the father of Henri Cartan.

Cartan, Henri (1904–2008) Noted complex analyst of the 20th century. Also a great humanitarian and teacher.

Castelnuovo, Guido (1865–1952) Noted Italian algebraic geometer of the early 20th century. Did important work on the classification of algebraic surfaces.

Cauchy, Augustin-Louis (1789–1857) A great analyst of the 19th century. Worked on both real and complex analysis. Is generally credited with making Newton's calculus rigorous.

Cayley, Arthur (1821–1895) A British pure mathematician. Put group theory on a modern footing. Invented matrix theory.

Chauvenet, William (1820–1870) Founder of the Naval Academy in Annapolis. Also founder of the Mathematics Department at Washington University in St. Louis. Served as chancellor of Washington University. Did the calculations for the Eads Bridge.

Chevalley, Claude (1909–1984) Noted American geometric analyst of the 20th century. Developer of modern algebraic geometry. Also a pioneer of Lie group theory.

Cohen, Paul J. (1924–2007) Noted American analyst. Prover of the independence of the continuum hypothesis, and winner of the Fields Medal.

Colmar, Thomas of (1785–1870) Inventor of the first genuine mechanical calculator.

Connes, Alain (1947–) Outstanding worker in the field of von Neumann algebras. Winner of the Fields Medal.

Coulomb, Jean (1904–1999) A member of the founding Bourbaki group in France.

Courant, Richard (1888–1972) Student of David Hilbert. Founder of the mathematics institute in Göttingen and also of the Courant Institute of the Mathematical Sciences at New York University. Influential 20th century analyst.

Cray, Seymour (1925–1996) The father of supercomputing and parallel processing.

Delsarte, Jean (1903–1968) One of the founding members of the Bourbaki group in France.

De Morgan, Augustus (1806–1871) Noted 19th century logician. Remembered for De Morgan's laws.

de Possel, René (1905–1974) One of the founding members of the Bourbaki group in France.

Dieudonné, Jean (1906–1992) Noted 20th century analyst and geometer. Mentor of Alexandre Grothendieck. One of the founding faculty of the Institut des Hautes Études Scientifiques in France.

Dirichlet, Peter Gustav Lejeune (1805–1859) One of the great number theorists of the 19th century. Creator of the pigeonhole principle. Teacher of Kronecker and Riemann.

Dubreil, Paul (1904–1994) One of the founding members of the Bourbaki group in France. He later dropped out.

Eckert, J. Presper (1919–1995) One of the inventors of the ENIAC computer at the University of Pennsylvania.

Ehresmann, Charles (1905–1979) One of the founding members of the Bourbaki group in France.

Enriques, Federigo (1871–1946) Noted Italian algebraic geometer of the early 20th century. One of the first to give a classification of algebraic surfaces in birational geometry.

Erdős, Paul (1913–1996) Noted Hungarian mathematician of the 20th century. Worked in all fields of mathematics. Never had a home institution or a regular home. Wrote 1500 papers. The inspiration for the concept of Erdős number.

Estermann, Theodor (1902–1991) Did important work on the Goldbach conjecture.

Euler, Leonhard (1707–1783) One of the great mathematicians of all time. His complete works comprise 70 volumes. Worked in all areas of mathematics, as well as physics, mechanics, and engineering.

Feller, William (1906–1970) One of the founding probabilists of the 20th century. A professor at Princeton University.

de Fermat, Pierre (1601–1665) The greatest amateur mathematician of all time. A noted magistrate in Toulouse, France. A celebrated number theorist, and father of Fermat's last theorem.

Feynman, Richard (1918–1988) Noted American physicist. Winner of the Nobel Prize. Father of the Feynman diagram and many other important ideas in quantum theory.

Fibonacci (Leonardo Pisano Bogollo) (1170–1250) A noted classical mathematician. Father of the Fibonacci series.

Fourier, Jean Baptiste Joseph (1768–1830) The father of Fourier series, and of the analytic theory of heat. An important mathematical physicist.

Franklin, Philip (1898–1965) Noted Fourier analyst. Protégé of Norbert Wiener.

Frege, Gottlob (1848–1925) Perhaps the greatest logician of the 19th century. Recipient of Bertrand Russell's famous letter about Russell's paradox.

Galois, Évariste (1812–1832)   Noted French mathematical genius who died very young. Inventor of Galois theory and group theory.

Garfield, James (1831–1881)   President of the United States, and also creator of one of the many proofs of the Pythagorean theorem.

Gauss, Carl Friedrich (1777–1855)   One of the three greatest mathematicians (along with Newton and Archimedes) who ever lived. Prover of the fundamental theorem of algebra. Creator of many seminal parts of number theory.

Gelbart, Abraham (1912–1994)   Noted mathematician who never got a high school diploma. Protégé of Norbert Wiener.

Germain, Sophie (1776–1831)   One of the greatest woman mathematicians of all time. Namesake of the "Sophie Germain primes." Proved a notable result about Fermat's last theorem.

Gilman, Daniel Coit (1831–1908)   Founding president of Johns Hopkins University.

Gödel, Kurt (1906–1978)   One of the great logician of all time. Prover of the famous Gödel incompleteness theorem. Inventor of recursive functions.

Goldbach, Christian (1690–1764)   Namesake of the Goldbach conjecture in number theory.

Goldfeld, Dorian (1947–)   Mathematician at Columbia University. Author of a history of the Erdős-Selberg dispute regarding the elementary proof of the prime number theorem.

Goldstine, Herman (1913–2004)   Co-inventor, with John von Neumann, of the first electronic stored-program computer.

Gorenstein, Daniel (1923–1992)   A major figure in the modern theory of finite groups. Organizer of the project to assemble the full classification of the finite simple groups.

Grothendieck, Alexandre (1928–)   One of the great mathematicians of the 20th century. Father of the theory of nuclear spaces. Winner of the Fields Medal. Father of the modern theory of algebraic geometry. Founding faculty member of the Institut des Hautes Études Scientifiques.

Hadamard, Jacques (1865–1963)   One of the great mathematical analysts of the past 150 years. Prover of the prime number theorem. Also a great humanitarian.

Halmos, Paul (1916–2006) Outstanding mathematical analyst of the 20th century. Noted mathematical expositor and teacher. Author of many famous mathematics texts.

Hamilton, William Rowan (1805–1865)   Noted British physicist, astronomer, and mathematician. Famous for his reformulation of Newton's mechanics. Namesake of the Hamiltonian.

Hardy, Godfrey H. (1877–1947) Great mathematical analyst of the 20th century. Collaborator of John E. Littlewood. Mentor of S. Ramanujan.

Heawood, Percy (1861–1955) Famous for finding the error in Kempe's proof of the four-color theorem and formulating the Heawood conjecture.

Heisenberg, Werner (1901–1976) The father of quantum mechanics. A great 20th century physicist.

Henry, Joseph (1797–1878) One of the founders of Johns Hopkins University.

Hilbert, David (1862–1943) One of the outstanding figures in 20th century mathematics. Is reputed to have been an expert in all parts of mathematics. Made seminal contributions to geometry, logic, algebra, and analysis. Was the leader of the mathematics group in Göttingen.

Hobbes, Thomas (1588–1679) Noted 16th century philosopher. Famous for attempting a mathematical theory of ethics.

Hollerith, Herman (1860–1929) Creator of an early computing machine. The father of International Business Machines (IBM).

Hooke, Robert (1635–1703) Noted 17th century mathematician and physicist. Friend of Isaac Newton. Namesake of Hooke's Law in the theory of springs. Hooke is noted for his theory of elasticity, and for coining the term "cell" (in botany).

Jacobi, Carl (1804–1851) Noted 19th century number theorist. Noted for his work in the theory of elliptic functions.

Jevons, William S. (1835–1882) Creator of the "logical piano" (an early computer).

Jobs, Steve (1955–) One of the founders of Apple Computer. Creator of the NEXT Computer, the IMac, the IPhone, and many other high-tech innovations.

Jordan, Camille (1838–1922) A great 19th century geometer. Namesake of the Jordan curve theorem. Also one of the provers of the classification theory of compact, two-dimensional surfaces.

Kaplansky, Irving (1917–2006) Noted 20th century algebraist. Made significant contributions to group theory, ring theory, operator algebras, and field theory. Served as chair of the University of Chicago Mathematics Department, and also as director of the Mathematical Sciences Research Institute.

Kelvin, Lord (William Thomson) (1900) A noted mathematician and physicist. Did important work on the first and second laws of thermodynamics and also in the theory of electricity.

Kempe, Alfred (1845–1922) Noted worker on the four-color theorem. Published a proof in 1879 that was discovered eleven years later to be false. The namesake of Kempe chains.

Kepler, Johannes (1571–1630) One of the great astronomers of all time. Student of Tycho Brahe. Formulator of Kepler's laws of planetary motion.

Kilburn, Tom (1921–2001) One of the developers of the Williams Tube, an early computer.

Klein, Felix (1849–1925) A great geometer of the 19th century. Colleague of David Hilbert. Namesake of the Klein bottle, and father of the Erlangen program in geometry.

Kolmogorov, Andrei (1903–1987) Distinguished 20th century analyst. Developer of the modern theory of probability. Did important work in Fourier series, stochastic processes, and turbulence. Had many distinguished Ph.D. students.

Kronecker, Leopold (1823–1891) Noted 19th century algebraist. A student of Dirichlet, and teacher of Cantor. Studied the solvability of equations and the continuity of functions.

Kummer, Ernst (1810–1893) Noted 19th century algebraist. Studied hypergeometic series and abelian varieties. Proved many cases of Fermat's last theorem.

Lagrange, Joseph-Louis (1736–1813) Noted geometric analyst of the 19th century. One of the fathers of the theory of celestial mechanics. Made important contributions to the calculus of variations, and invented Lagrange multipliers.

Lebesgue, Henri (1875–1941) The father of measure theory. An important analyst of the 20th century.

Leibniz, Gottfried Wilhelm von (1646–1716) An important philosopher. One of the inventors of calculus. The author of *The Monadology*.

Leray, Jean (1906–1998) An important geometer and algebraist of the 20th century. The inventor of sheaves, and also of spectral sequences.

Levinson, Norman (1912–1975)   A protégé of Norber Wiener. A distinguished mathe-matical analyst, and professor at MIT. Proved a significant result about the zeros of the Riemann zeta function.

Lobachevsky, Nikolai (1793–1856)   Along with Bolyai, one of the inventors of non-Euclidean geometry.

Mandelbrot, Benoît (1924–2010)   Father of fractal theory. Namesake of the Mandelbrot set. A great expositor, and exponent of the idea of fractional dimension and fractal dimension.

Mandelbrojt, Szolem (1899–1983)   Distinguished mathematical analyst of the 20th century. Uncle of Benoît Mandelbrot. Studied Fourier series.

Mauchly, John W. (1907–1980)   One of the creators of the ENIAC computer at the University of Pennsylvania.

Mittag-Leffler, Gösta (1846–1927)   Distinguished Swedish mathematician of the late 19th and early 20th centuries. Founder of the Mittag-Leffler Institute. Student of Weierstrass. Possible reason that mathematicians do not receive the Nobel Prize.

Möbius, August (1790–1868)   Noted 19th century geometer. Namesake of the Möbius strip. Prover of the result classifying all two-dimensional surfaces.

Morse, Marston (1892–1977)   Distinguished 20th century geometric analyst. The name-sake of Morse theory, or the calculus of variations in the large. Permanent member of the Institute for Advanced Study.

Napier, John (1550–1617)   The inventor of logarithms.

von Neumann, John (1903–1957)   Noted American mathematical analyst of Hungarian origin. John von Neumann created the modern view of quantum mechanics, and is the father of the subject of von Neumann algebras. He made notable contributions to functional analysis, logic, and many other parts of mathematics. He is considered to be the father of the stored-program computer.

Nevanlinna, Rolf (1895–1980)   Distinguished 20th century complex analyst. Teacher of Lars Ahlfors. Namesake of Nevanlinna theory.

Newell, Allen (1927–1992)   One of the creators of the Logic Theory Machine, an early device for discovering and creating mathematical proofs.

Newson, Mary Winston (1869–1959)   The first American woman to earn a Ph.D. degree at the University of Göttingen. She produced the English translation of the transcription of Hilbert's 23 problems delivered at the International Congress of Mathematicians in 1900.

Nobel, Alfred (1833–1896)   Inventor of dynamite. Wealthy industrialist. Founder of the Nobel Prize.

Occam, William of (1288–1348)   Fourteenth century philosopher. Creator of "Occam's razor," the precept that any logical system should be as compact and elegant as possible.

Ohm, Georg (1787–1854)   Namesake of Ohm's law. Studied electric current. Also worked on the development of the battery. Published works on acoustics.

Painlevé, Paul (1863–1933)   A distinguished 19th- and 20th century analyst. Professor at the Sorbonne. Studied differential equations, gravitation, and complex analysis. Also was involved in politics.

Pascal, Blaise (1623–1662)   Noted 17th century French philosopher. One of the founders of probability theory. The namesake of Pascal's triangle. Author of *Pensées*.

Peirce, Benjamin (1809–1880)   Professor of mathematics at Harvard University. Namesake of the Peirce Instructorships. Worked on celestial mechanics, number theory, algebra, and the philosophy of mathematics.

Poincaré, Henri (1854–1912)   One of the great mathematicians of the 20th century. The creator of dynamical systems theory. One of the fathers of topology. The namesake of the Poincaré conjecture—recently proved by Grigori Perelman.

Prawitz, D. (1936–)   Creator of one of the early machine-driven theorem-proving technologies.

Rademacher, Hans (1892–1969)   Distinguished analyst and number theorist at the University of Pennsylvania. In 1945 he announced that he could disprove the Riemann hypothesis—but he was mistaken.

Ramanujan, Srinivasa (1887–1920)   Great mathematical genius from India discovered almost by accident by G. H. Hardy. Worked with Hardy at Cambridge University, mainly in number theory. Died young of ill health.

Riemann, Bernhard (1826–1866)   Great mathematical analyst of the 19th century. Made seminal contributions to calculus, real analysis, complex analysis, and geometry. Created Riemann surfaces, and created the geometry that was important for Einstein's general theory of relativity. Died young of ill health.

Rutherford, Ernest (1871–1937)   Distinguished British mathematical physicist. Considered to be the father of nuclear physics. Credited with splitting the atom. Won the Nobel Prize in Chemistry in 1908. Noted for being a skeptic of the theory of relativity.

Schrödinger, Erwin (1887–1961) One of the fathers of quantum theory. A distinguished Austrian physicist. Famous for Schrödinger's wave equation and "Schrödinger's cat."

Selberg, Atle (1917–2007) Distinguished Norwegian analytic number theorist. Famous for results on the zeros of the Riemann zeta function, for the elementary proof of the prime number theorem, and also for the Selberg trace formula.

Serre, Jean-Pierre (1926–) One of the great algebraists and geometers of the 20th century. Made seminal contributions to algebraic geometry, algebraic topology, number theory, and to many other fields. Winner of the Fields Medal, the Abel Prize, and many other encomia.

Simon, Herbert A. (1916–2001) One of the creators of the Logic Theory Machine, an early machine that could discover and create mathematical proofs.

Steinhaus, Hugo (1887–1972) Distinguished 20th century analyst. Mathematical grandfather of Antoni Zygmund. Discoverer of the uniform boundedness principle and important ideas in the theory of Fourier series.

Stone, Marshall (1903–1989) Distinguished American analyst. Namesake of the Stone–Weierstrass theorem. Important chair of the mathematics department at the University of Chicago.

Sylvester, James J. (1814–1897) Distinguished 19th century algebraist. Played an important role in developing mathematics at Johns Hopkins University, and in America at large. Founder of the *American Journal of Mathematics*.

Tarski, Alfred (1902–1983) Outstanding Polish logician. On the faculty at the University of California at Berkeley for many years, he played a dynamic role in developing modern logic. Famous for the Banach–Tarski paradox.

Turing, Alan (1912–1954) Widely considered to be one of the great mathematical geniuses of the 20th century. He invented the Turing machine, and had seminal ideas in the design of early computers. Did important cryptographic work in World War II. Died of poisoning at an early age.

Vallée Poussin, Charles Jean Gustav Nicolas Baron de la (1866–1962) Outstanding analyst of the late 19th and early 20th centuries. One of the provers of the prime number theorem. One of the first and only winners of the Mittag-Leffler Medal.

Vinogradov, Ivan M. (1891–1983) One of the fathers of modern analytic number theory. Famous for his analysis of exponential sums. He served as director of the Steklov Mathematics Institute for 49 years, and was a distinguished leader of Soviet mathematics.

von Kármán, Theodore (1881–1963)   A student of David Hilbert, von Kármán is considered to have been the founder of modern aeronautical engineering. Did important work in supersonic and hypersonic airflow.

vos Savant, Marilyn (1946–)   Originally born Marilyn Mach, she is a popular newspaper columnist whose claim to fame is that she is reputed to have the highest IQ in the world. She is married to Robert K. Jarvik, who invented the artificial heart.

van der Waerden, Bartel L. (1903–1996)   Important Dutch algebraist. Student of Emmy Noether. Wrote a definitive text in algebra. Proved an important result in Ramsey theory.

Wang, Hao (1921–1995)   A developer of one of the earlist theorem-proving machines.

Weierstrass, Karl (1815–1897)   A distinguished German analyst of the 19th century. Weierstrass made seminal contributions to both real and complex analysis. He is particularly noted for some important examples. His theorems are also frequently quoted.

Weil, André (1906–1998)   Widely considered to be one of the great mathematicians of the 20th century. He was a permanent member of the Institute for Advanced Study. Weil made seminal contributions to algebraic geometry, number theory, invariant integrals, Kähler manifold theory, and many other parts of mathematics. His books have been very influential.

Weinberg, Steven (1933–)   Noted American physicist. Winner of the Nobel Prize for studying the unification of the weak force and electromagnetic interaction between elementary particles.

Weyl, Hermann (1885–1955)   Noted German mathematician of the 20th century. Particularly respected for his work in algebra and invariant theory. A permanent member of the Institute for Advanced Study.

Whitney, Hassler (1907–1989)   A noted geometric analyst of the 20th century. Proved important results about manifolds and about extension of functions. A permanent member of the Institute for Advanced Study.

Wiener, Norbert (1894–1964)   Noted American analyst. A student of Littlewood. Faculty member at MIT for many years. Made seminal contributions to Fourier analysis, stochastic processes. Invented the concept of cybernetics.

Williams, Frederic (1911–1977)   Collaborated in developing the early computer known as the Williams Tube.

Wolfram, Stephen (1959–)   Young prodigy in mathematics and physics. Youngest recipient ever of the MacArthur Fellowship. Inventor of the software `Mathematica`. Author of the book *A New Kind of Science*.

Wozniak, Steve (1950–)   Cofounder, along with Steve Jobs, of Apple Computer. Inventor of the $256K$ memory chip. Wozniak was the "technical talent" in getting Apple Computer off the ground.

Zariski, Oskar (1899–1986)   One of the great algebraic geometers of the 20th century. Namesake of the Zariski topology and inventor of the concept of "blowup" (which led to Hironaka's Fields-Medal-winning work on resolution of singularities). Zariski was the Ph.D. advisor of two Fields Medalists—David Mumford and Heisuke Hironaka.

Zuse, Konrad (1910–1995)   Creator of the first freely programmable computer, called the Z1.

References

[ADL1]   A. Adler, The second fundamental forms of S^6 and $P^n(C)$, *Am. J. Math.* 91(1969), 657–670.

[ADL2]   A. Adler, Mathematics and creativity, The *New Yorker* 47(1972), 39–45.

[AKS]   M. Agrawal, N. Kayal and N. Saxena, PRIMES is in P, *Annals of Math.* 160(2004), 781–793.

[ALM]   F. J. Almgren, *Almgren's Big Regularity Paper. Q-Valued Functions Minimizing Dirichlet's Integral and the Regularity of Area-Minimizing Rectifiable Currents up to Codimension 2*, World Scientific Publishing Company, River Edge, NJ, 1200.

[APH1]   K. Appel and W. Haken, A proof of the four color theorem, *Discrete Math.* 16(1976), 179–180.

[APH2]   K. Appel and W. Haken, Every planar map is four colorable, *Illinois J. Mathematics* 21(1977), 429–567.

[APH3]   K. Appel and W. Haken, The four color proof suffices, *Math. Intelligencer* 8(1986), 10–20.

[APH4]   K. Appel and W. Haken, *Every Planar Map Is Four Colorable*, American Mathematical Society, Providence, RI, 1989.

[ASC]   M. Aschbacher, Highly complex proofs and implications of such proofs, *Phil. Trans. R. Soc. A* 363(2005), 2401–2406.

[ASM]   M. Aschbacher and S. Smith, *The Classification of Quasithin Groups*, I and II, American Mathematical Society, Providence, RI, 2004.

[ASW]   T. Aste and D. Weaire, *The Pursuit of Perfect Packing*, Institute of Physics Publishing, Bristol, U.K., 2000.

[ATI1]   M. Atiyah, Responses to Jaffe and Quinn 1994, *Bulletin of the AMS* 30(1994), 178–207.

[ATI2]   M. Atiyah, Mathematics: Art and Science, *Bulletin of the AMS* 43(2006), 87–88.

[AVI]   J. Avigad, Proof mining, notes from ASL 2004, `http://www.andrew.cmu.edu/avigad`.

[BAB1]   D. H. Bailey and J. M. Borwein, *Mathematics by Experiment*, A K Peters Publishing, Natick, MA, 2004.

[BAB2]   D. H. Bailey and J. M. Borwein, Experimental mathematics: examples, methods and implications, *Notices of the AMS* 52(2005), 502–514.

[BAB3]   D. H. Bailey and J. M. Borwein, Computer-assisted discovery and proof, preprint.

[GAR]   M. Gardner, *The Second Scientific American Book of Mathematical Puzzles and Diversions*, University of Chicago Press, Chicago, IL, 1987.

[GAJ]   M. R. Garey and D. S. Johnson, *Computers and Intractability: A Guide to the Theory of NP-Completeness*, W. H. Freeman and Company, San Francisco, CA, 1991.

[GON]   G. Gonthier, Formal proof—the four-color theorem, *Notices of the American Mathematical Society* 55(2008), 1382–1393.

[GLS]   D. Gorenstein, R. Lyons and R. Solomon, *The Classification of the Finite Simple Groups*, American Mathematical Society, Providence, RI, 1994.

[GRA]   J. Gray, *The Hilbert Challenge*, Oxford University Press, New York, 2000.

[GRT]   B. Green and T. Tao, The primes contain arbitrarily long arithmetic progressions, *Annals of Math.* 167(2008), 481–547.

[GRE]   M. J. Greenberg, *Euclidean and Non-Euclidean Geometries*, 2nd ed., W. H. Freeman, New York, 1980.

[GRE]   B. Greene, *The Elegant Universe*, W. W. Norton & Co., New York, 2003.

[HAL1]   T. Hales, The status of the Kepler conjecture, *Math. Intelligencer* 16(1994), 47–58.

[HAL2]   T. Hales, A proof of the Kepler conjecture, *Annals of Math.* 162(2005), 1065–1185.

[HAL3]   T. Hales, Formal proof, *Notices of the American Mathematical Society* 55(2008), 1370–1380.

[HAL]   M. Hall, *The Theory of Groups*, Macmillan, New York, 1959.

[HAR]   G. H. Hardy, *A Mathematician's Apology*, Cambridge University Press, London, 1967.

[HER]   I. M. Herstein, *Topics in Algebra*, Xerox, Lexington, 1975.

[HIN]   G. Higman and B. H. Neumann, Groups as groupoids with one law, *Publicationes Mathematicae Debreceu* 2(1952), 215–227.

[HIL]   D. Hilbert, *Grundlagen der Geometrie*, Teubner, Leipzig, 1899.

[HIA]   D. Hilbert and W. Ackermann, *Grundzüge der theoretischen Logik*, Springer-Verlag, Berlin, 1928.

[HHM]   D. Hoffman, The computer-aided discovery of new embedded minimal surfaces, *Math. Intelligencer* 9(1987), 8–21.

[HOF]   P. Hoffman, *The Man Who Loved Only Numbers: The Story of Paul Erdős and the Search for Mathematical Truth*, Hyperion, New York, 1999.

[HOR1]   J. Horgan, The Death of Proof?, *Scientific American* 269(1993), 93–103.

[HOR2]   J. Horgan, *The End of Science*, Broadway Publishers, New York, 1997.

[HOS]   E. Horowitz and S. Sahni, Exact and approximate algorithms for scheduling nonidentical processors, *J. Assoc. Comput. Mach.* 23(1976), 317–327.

[HRJ]   K. Hrbacek and T. Jech, *Introduction to Set Theory*, 3rd ed., Marcel Dekker, New York, 1999.

[HSI1]   W.-Y. Hsiang, On the sphere packing problem and the proof of Kepler's conjecture, *Internat. J. Math.* 4(1993), 739–831.

[HSI2]   W.-Y. Hsiang, Sphere packings and spherical geometry—Kepler's conjecture and beyond, Center for Pure and Applied Mathematics, U. C. Berkeley, July, 1991.

[HSI3]   W.-Y. Hsiang, *Least Action Principle of Crystal Formation of Dense Packing Type and Kepler's Conjecture*, World Scientific, River Edge, NJ, 2001.

[HSI4]   W.-Y. Hsiang, A rejoinder to T. C. Hales's article: "The status of the Kepler conjecture" *Math. Intelligencer* 16(1994), *Math. Intelligencer* 17(1995), 35–42.

[IFR]   G. Ifrah, *The Universal History of Numbers: From Prehistory to the Invention of Computers*, translated from the French by David Bellos, E. F. Harding, Sophie Wood, and Ian Monk, John Wiley & Sons, New York, 1998.

[JAF]   Arthur Jaffe, Proof and the evolution of mathematics, *Synthese* 111(1997), 133–146.

[JAQ]   Arthur Jaffe and F. Quinn, "Theoretical mathematics": toward a cultural synthesis of mathematics and theoretical physics, *Bulletin of the A.M.S.* 29(1993), 1–13.

[JEC]   T. Jech, *The Axiom of Choice*, North-Holland, Amsterdam, 1973.

[KAP1]   R. Kaplan and E. Kaplan, *The Nothing That Is: A Natural History of Zero*, Oxford University Press, Oxford, 2000.

[KAP2]   R. Kaplan and E. Kaplan, *The Art of the Infinite: The Pleasures of Mathematics*, Oxford University Press, Oxford, 2003.

[KAR]   R. M. Karp, Reducibility among combinatorial problems, in R. E. Miller and J. W. Thatcher, eds., *Complexity of Computer Computations*, Plenum Press, New York, 1972, 85–103.

[KLL]   B. Kleiner and J. Lott, Notes on Perelman's papers, `arXiv:math.DG/0605667`.

[KLN]   M. Kline, *Mathematics, The Loss of Certainty*, Oxford University Press, New York, 1980.

[KLI1]   W. Klingenberg, Existence of infinitely many closed geodesics, *J. Differential Geometry* 11(1976), 299–308.

[KLI2]   W. Klingenberg, *Lectures on closed geodesics*, 3rd ed., Mathematisches Institut der Universität Bonn, Bonn, 1977.

[KLI3]   W. Klingenberg, *Lectures on Closed Geodesics*, Grundlehren der Mathematischen Wissenschaften, v. 230, Springer-Verlag, Berlin-New York, 1978.

[KOL]   G. Kolata, Mathematical proof: The genesis of reasonable doubt, *Science* 192(1976), 989–990.

[KRA1]   S. G. Krantz, The immortality of proof, *Notices of the AMS*, 41(1994), 10–13.

[KRA2]   S. G. Krantz, *A Primer of Mathematical Writing*, American Mathematical Society, Providence, RI, 1996.

[KRA3]   S. G. Krantz, *Mathematical Publishing: A Guidebook*, American Mathematical Society, Providence, RI, 2005.

[KRA4]   S. G. Krantz, *Handbook of Logic and Proof Techniques for Computer Science*, Birkhäuser, Boston, 2002.

[KRA5]   S. G. Krantz, *The Elements of Advanced Mathematics*, 2nd ed., CRC Press, Boca Raton, FL, 2002.

[KRA6]   S. G. Krantz, Review of *A New Kind of Science*, *Bull. AMS* 40(143–150).

[KRA7]   S. G. Krantz, *Real Analysis and Foundations*, Taylor & Francis, Boca Raton, FL, 2005.

[KUH]   T. S. Kuhn, *The Structure of Scientific Revolutions*, 2nd ed., University of Chicago Press, Chicago, IL, 1970.

[KUN]   K. Kunen, Single axioms for groups, *J. Automated Reasoning* 9(1992), 291–308.

[LAK]   I. Lakatos, *Proofs and Refutations*, Cambridge University Press, Cambridge, 1976.

[LAW]   E. L. Lawler, A pseudopolynomial algorithm for sequencing jobs to minimize total tardiness, *Ann. Discrete Math.* 1(1977), 331–342.

[LEC]   M. Lecat, *Erreurs de Mathématiciens des origines à nos jours*, Ancienne Libraire Castaigne et Librairie *Ém* Desbarax, Brussels, 1935.

[LRKF]   J. K. Lenstra, A. H. G. Rinnooy Kan and M. Florian, Deterministic production planning: algorithms and complexity, unpublished manuscript, 1978.

[XJLI]   X.-J. Li, A proof of the Riemann hypothesis, `arXiv:0807.0090`.

[LIT]   J. E. Littlewood, *A Mathematician's Miscellany*, Methuen, London, 1953.

[LOV]   L. Lovasz, Coverings and colorings of hypergraphs, *Proceedings of the 4th Southeastern Conference on Combinatorics, Graph Theory, and Computing*, Utilitas Mathematica Publishing, Winnipeg, 1973, 3–12.

[MCI]   A. MacIntyre, The mathematical significance of proof theory, *Phil. Trans. R. Soc.* A 363(2005), 2419–2435.

[MAK]   D. Mackenzie, The automation of proof: a historical and sociological exploration, *IEEE Annals of the History of Computing* 17(1995), 7–29.

[MAC]   S. Mac Lane, Mathematical models, *Am. Math. Monthly* 88(1981), 462–472.

[MAN1]   B. Mandelbrot, *The Fractal Geometry of Nature*, Freeman, New York, 1977.

[MAN2]   B. Mandelbrot, Responses to "Theoretical mathematics: toward a cultural synthesis of mathematics and theoretical physics," by A. Jaffe and F. Quinn, *Bulletin of the AMS* 30(1994), 193–196.

[MAA]   K. Manders and L. Adleman, NP-complete decision problems for binary quadratics, *J. Comput. System Sci.* 16(1978), 168–184.

[MAN]   A. L. Mann, A complete proof of the Robbins conjecture, preprint.

[MAZ]   B. Mazur, Mathematical Platonism and its opposites, `http://www.math.harvard.edu/~mazur/`.

[MCC]   J. McCarthy, *Review of The Emperor's New Mind by Roger Penrose*, *Bull. AMS* 23(1990), 606–616.

[MCU]   W. McCune, Single axioms for groups and abelian groups with various operations, *J. Automated Reasoning* 10(1993), 1–13.

[MOT]   J. W. Morgan and G. Tian, *Ricci Flow and the Poincaré Conjecture*, The Clay Institute of Mathematics, American Mathematical Society, Providence, RI, 2007,

[NAG]   S. Nasar and D. Gruber, Manifold Destiny, *The New Yorker*, August 28, 2006.

[NCBI]   collection of articles about proof, *PubMed*,
```
http://www.ncbi.nlm.nih.gov/sites/entrez?Db
=pubmed&DbFrom=pubmed&Cmd=Link&LinkName
=pubmed_pubmed&LinkReadableName
=Related%20Articles&IdsFromResult
=16188617&ordinalpos=1&itool
=EntrezSystem2.PEntrez.Pubmed.Pubmed_ResultsPanel
.Pubmed_DiscoveryPanel.Pubmed_Discovery_RA
&log$=relatedarticles&logdbfrom=pubmed.
```

[OKT]   S. Oka and H. Toda, Non-triviality of an element in the stable homotopy groups of spheres, *Hiroshima Math. J.* 5(1975), 115–125.

[PEN1]   R. Penrose, *The Emperor's New Mind: Concerning Computers, Minds, and the Laws of Physics*, Oxford University Press, Oxford, 1989.

[PEN2]   R. Penrose, *The Road to Reality: A Complete Guide to the Laws of the Universe*, Jonathan Cape, London, 2004.

[PER1]   G. Perelman, The entropy formula for the Ricci flow and its geometric applications, `arXiv:math.DG/0211159v1`.

[PER2]   G. Perelman, Ricci flow with surgery on three-manifolds, `arXiv:math.DG/0303109v1`.

[PER3]   G. Perelman, Finite extinction time for the solutions to the Ricci flow on certain three-manifolds, `arXiv:math.DG/0307245v1`.

[PTRS]   Table of Contents, *Phil. Trans. R. Soc.* A, v. 363 n. 1835(2005),
```
http://journals.royalsociety.org/content/
j2716w637777/?p=e1e83758d3ee4a51b0aa0bba4133518a
&pi=0.
```

[POP]   K. R. Popper, *Conjectures and Refutations: The Growth of Scientific Knowledge*, Basic Books, New York, 1962.

[RIE]   B. Riemann, On the number of primes less than a given magnitude, *Monthly Reports of the Berlin Academy*, 1859.

[RSST]   N. Robertson, D. P. Sanders, P. D. Seymour and R. Thomas, A new proof of the four-color theorem, *Electr. Res. Announc. AMS* 2(1996), 17–25.

[ROB]   A. Robinson, *Nonstandard Analysis*, North Holland, Amsterdam, 1966.

[RUD]   W. Rudin, *Principles of Real Analysis*, 3rd ed., McGraw-Hill, New York, 1976.

[RUS]   B. Russell, *History of Western Philosophy*, Routledge, London, 2004.

[SAB]   Karl Sabbagh, *The Riemann Hypothesis: The Greatest Unsolved Problem in Mathematics*, Farrar, Straus, & Giroux, New York, 2003.

[SAV]   M. vos Savant, *The World's Most Famous Math Problem*, St. Martin's Press, New York, 1993.

[SCHA]   T. J. Schaefer, Complexity of some two-person perfect-information games, *J. Comput. Syst. Sci.* 16(1978), 185–225.

[SCL]   C. P. Schnorr and H. W. Lenstra, Jr., A Monte Carlo factoring algorithm with linear storage, *Math. Comput.* 43(1984), 289–311.

[SEY]   P. Seymour, Progress on the four-color theorem, *Proceedings of the ICM* (Zürich, 1994), 183–195, Birkhäuser, Basel, 1995.

[SMA]   S. Smale, Review of E. C. Zeeman: Catastrophe Theory, Selected Papers 1972–1977, *Bulletin of the AMS* 84(1978), 1360–1368.

[SMU1]   R. Smullyan, *Forever Undecided: A Puzzle Guide to Gödel*, Alfred Knopf, New York, 1987.

[SMU2]   R. Smullyan, *The Lady or the Tiger? and Other Logic Puzzles*, Times Books, New York, 1992.

[STA1]   St. Andrews Biography of al-Khwarizmi, `http://www-history.mcs.st-andrews.ac.uk/Biographies/Al-Khwarizmi.html`.

[STA2]   St. Andrews Biography of Fermat, `http://www-hostory.mcs.st-andrews.ac.uk/Biographies/Fermat.html`.

[STI]   M. Stickel, A case study of theorem proving by the Knuth-Bendix method: discovering that $x^3 = x$ implies ring commutativity, *Proceedings of the Seventh International Conference on Automated Deduction*, R. E. Shostak, ed., Springer-Verlag, New York, 1984, pp. 248–258.

[STM]   L. J. Stockmeyer and A. R. Meyer, Word problems requiring exponential time, *Proceedings of the 5th Annual ACM Symposium on Theory of Computing*, Association for Computing Machinery, New York, 1973, 1–9.

[STR]   R. S. Strichartz, letter to the editor, *Notices of the American Mathematical Society* 53(2006), 406.

[STR1]   W. R. Stromquist, Some Aspects of the Four Color Problem, Ph.D. thesis, Harvard University, 1975.

[STR2]   W. R. Stromquist, The four-color theorem for small maps, *J. Combinatorial Theory* 19(1975), 256–268.

[STRZ]   Pawel Strzelecki, The Poincaré conjecture?, *American Mathematical Monthly* 113(2006), 75–78.

[SUP]   P. Suppes, *Axiomatic Set Theory*, Van Nostrand, Princeton, 1972.

[SWD]   P. Swinnerton-Dyer, The justification of mathematical statements, *Phil. Trans. R. Soc.* A 363(2005), 2437–2447.

[THZ]   E. Thomas and R. Zahler, Nontriviality of the stable homotopy element γ_1, *J. Pure Appl. Algebra* 4(1974), 189–203.

[THU1]   W. P. Thurston, 3-dimensional manifolds, Kleinian groups and hyperbolic geometry, *Bull. AMS* 6:3(1982), 357–381.

[THU2]   W. P. Thurston, On proof and progress in mathematics, *Bull. AMS* 30(1994), 161–177.

[THU3]   W. P. Thurston, *3-dimensional Geometry and Topology*, Vol. 1, Princeton University Press, Princeton, 1997.

[THU3]   W. P. Thurston, *The Geometry and Topology of Three-Manifolds*, notes, Princeton University, 1980, 502 pp.

[WAG]   S. Wagon, *The Banach–Tarski Paradox*, Cambridge University Press, Cambridge and New York, 1985.

[WEIL1]   A. Weil, *Basic Number Theory*, 2nd ed., Springer-Verlag, New York, 1973.

[WEIL2]   A. Weil, *The Apprenticeship of a Mathematician*, Birkhäuser, Basel, 1992.

[WEI]   L. Weinstein, The Bieberbach conjecture, *International Math. Res. Notices* 5(1991), 61–64.

[WIG]   E. Wigner, The unreasonable effectiveness of mathematics in the natural sciences, *Comm. Pure App. Math.* 13(1960), 1–14.

[WRU]   A. N. Whitehead and B. Russell, *Principia Mathematica*, Cambridge University Press, Cambridge, 1910.

[WIE]   F. Wiedijk, Formal proof—getting started, *Notices of the American Mathematical Society* 55(2008), 1408–1414.

[WIL]   A. Wiles, Modular elliptic curves and Fermat's last theorem, *Annals of Math.* 141(1995), 443–551.

[WILS]   L. Wilson, *The Academic Man, A Study in the Sociology of a Profession*, Oxford University Press, London, 1942.

[WOLF]   R. S. Wolf, *A Tour Through Mathematical Logic*, A Carus Monograph of the Mathematical Association of America, Washington, D.C., 2005.

[WOL]   S. Wolfram, *A New Kind of Science*, Wolfram Media, Inc., Champaign, IL, 2002.

[WOS1]   L. Wos, *Automated Reasoning: Introduction and Applications*, Prentice-Hall, Englewood Cliffs, NJ, 1984.

[WOS2]   L. Wos, *Automation of Reasoning: An Experimenter's Notebook with Otter Tutorial*, Academic Press, New York, 1996.

[WOS3]   L. Wos, Automated reasoning answers open questions, *Notices of the AMS* 40(1993), 15–26.

[YAN]   B. H. Yandell, *The Honors Class*, A. K. Peters, Natick, MA, 2002.

[YEH]   R. Yehuda, Why do we prove theorems?, *Philos. Math.* 7(1999), 5–41.

[ZAH]   R. Zahler, Existence of the stable homotopy family, *Bull. Amer. Math. Soc.* 79(1973), 787–789.

[ZEE]   E. C. Zeeman, Controversy in science: on the ideas of Daniel Bernoulli and René Thom, The 1992/3 Johann Bernoulli Lecture, Gröningen, *Nieuw Archief van de Wiskunde* 11(1993), 257–282.

[ZEI]   D. Zeilberger, Theorems for a price: Tomorrow's semi-rigorous mathematical culture, *Notices of the AMS* 40(1993), 978–981.

Index